BIOLOGY EVERYWHERE

HOW THE SCIENCE OF LIFE MATTERS
TO EVERYDAY LIFE

Melanie E. Peffer, PhD

MKPEF4
Greeley, CO

TESTIMONIALS FOR
BIOLOGY EVERYWHERE

Biology Everywhere takes a unique approach to connecting biology content with common experiences most people will likely share. The content is easy to understand and takes the reader on a fascinating personal journey into the sometimes unexpected ways we interact with biology on a daily basis.

-Dr. Kristy Daniel, Associate Professor
Department of Biology, Texas State University

I love how Dr. Peffer integrates science with common, everyday topics of interest to the average person. She really accomplishes her goal of making science less academic and scary and much more approachable and digestible. Scientific literacy is one of the most important issues for our society, and this book will go a long way in addressing that important topic. I will absolutely be recommending this book for my students and their families.

-Mandy Revak, KidScience and ZooU Coordinator
Pittsburgh Zoo and PPG Aquarium

Biology Everywhere shows the reader that biology is quite the opposite of difficult to understand. Filled with relatable stories from the history of biology, along with personal and funny anecdotes, the book makes biology easy, directly relevant, and accessible to anyone. Packed with information, Dr. Peffer takes questions and scenarios we encounter in daily life and turns them into the fodder for biological explanations, never failing to draw the reader in. Anyone can read and learn from this book, from young adults to experts in biology.

-Dr. Jennifer Knight, Associate Professor
Department of Molecular, Cellular, and Developmental Biology
University of Colorado Boulder

Copyright © 2020 by Melanie E. Peffer, PhD.

All rights reserved. No part of this book may be reproduced in any written, electronic, recording, or photocopying form without written permission of the publisher or author. The exception would be in the case of brief quotations embodied in articles or reviews and pages where permission is specifically granted by the publisher or author.

Biology Everywhere/MKPEF4
Printed in the United States of America

Although every precaution has been taken to verify the accuracy of the information contained herein, the author and publisher assume no responsibility for any errors or omissions. No liability is assumed for damages that may result from the use of information contained within.

Biology Everywhere/Melanie E. Peffer, PhD -- 1st ed.

LCCN: 2020902960

ISBN Hardcover Edition: 978-1-7346531-0-6
ISBN Softcover Edition: 978-1-7346531-2-0
ISBN Ebook Edition: 978-1-7346531-1-3

CONTENTS

1. Why Biology Everywhere?.. 1

2. …because SCIENCE.. 7
   a. There is no universal scientific method
   b. Science is done by groups of people
   c. Science is by no means an unbiased process
   d. Creativity belongs to science as well as to the arts, and the best scientists are the creative ones, not the ones who are always right
   e. Knowledge generated by scientists changes in light of new evidence, and therefore it is not possible to prove anything definitively
   f. Hypotheses become theories and theories become laws, right?

3. Chemistry, Essentials Oils, and Venom 25
   a. So, what is a chemical?
   b. Water is a chemical and essential for life
   c. Natural and synthetic chemicals
   d. The chemistry of eating

4. Cells Gone Wild: Cancer 37
   a. Basics of cell biology
   b. What is cancer?
   c. How do people get cancer?
   d. What makes cells go rogue?

5. Energy: From the Sun to Your Granola Bar................................. 49
   a. Transfer of energy
   b. Glucose and fueling our bodies
   c. How do our cells use glucose, and what does it have to do with beer?

6. Mom Genes: DNA, Genetics, and Parenthood 61
    a. Rosalind Franklin and DNA
    b. Real-life X-Men: DNA mutations, sex, and you
    c. Mendel was lucky
    d. What about the epigenome?

7. Oh, Brave New Word 79
    a. Gene sequencing: Empowering good health decisions, or scaring the pants off everyone?
    b. Frankenfoods, or just speeding things along?
    c. A quick foray into really exciting, yet ethically dubious, technologies
        i. Post-implantation genetic diagnosis (PGD)
        ii. The science of de-extinction
        iii. Three-parent babies

8. The Diversity of Life: A Foray Into Seashells 95
    a. The "E" word
    b. Evolution produces new species
    c. "What's this? What's this?"
    d. Biodiversity: What it Means and Why We Need It

9. The Interconnectedness of Life and Issues of Conservation 115
    a. What is ecology?
    b. Questions of conservation: Finding balance
        i. The bushmeat crisis: Save the [insert endangered species here], or feed starving humans?
        ii. Is my reusable grocery bag actually greener?
        iii. Composting and organic waste
    c. Ecological crossroads and Hardin's Tragedy of the Commons

10. Tiny Humans: An Expose at the Intersection of Psychology and Biology 133
    a. In the beginning …
    b. Pregnancy and No No lists

    c. Congratulations, you have a tiny human! Now what?
    d. Childhood
    e. Adolescence and emerging adulthood
    f. The fountain of youth

11. The Psychology of Using Biological Evidence to Make Decisions 151
    a. Epistemological beliefs about science
    b. Certainty of science knowledge
    c. Structure of science knowledge
    d. Justification of science knowledge
    e. Source of science knowledge

12. Everything I Needed to Know about Life, I Learned in Band: The Arts and Biology 171
    a. Art as foundational to biology
    b. Physiology of music
        i. Listening to music
        ii. Making music
    c. The STEAM movement: Incorporating arts in science education

13. Finding New Treatments: The Business of Biology 187
    a. How does biological research work?
    b. Funding priorities: What gets funded and why?
    c. Applying basic biological research: Human subjects research and clinical trials

14. Next steps and Big Ideas in Biology Education 203
    a. Importance of quality biology education: Why does it matter?
    b. It's not just the pedagogy, attitudes are important, too
    c. Closing thoughts

To Felix, the light of my life.

ACKNOWLEDGEMENTS

I am enormously grateful for the support of so many friends, family members, students, and colleagues in the creation of Biology Everywhere. First off, I'd like to thank my husband, Kevin, for his unwavering support of this project and of my career and for listening to all of my biology everywhere anecdotes over the last 15 years. I also want to thank him for reviewing and editing so many things I've written, from high school biology essays to what I've written now. Thank you to Franklin Taggart at the Larimer Small Business Development Center for initially suggesting I turn my ideas into a book. I am indebted to all of my students, in particular the students who took non-STEM majors biology with me, for the candid discussions about their prior experiences in science classes and what works for them, and for all of the students who reached out at the end of the semester to say that it was my class that made all of the difference to changing their mind about science and biology. Thank you for giving me the confidence to share this approach of teaching biology to the world. I'm grateful to all of my colleagues who spent time sharing resources, reading drafts, and advising me on the content of this book. A special thank you goes to Kalyn Garcia (Kendall Reagan Nutrition Center), Dr. Jonathan Weinbaum (Southern Connecticut State University), Dr. Amy Keagy (University of North Florida), Dr. Sara Adkins (University of Alabama at Birmingham), Mandy Revak (Pittsburgh Zoo and PPG Aquarium), Dr. Shirley Smithson, Dr. Jennifer Knight (University of Colorado), Dr. Tom McCabe (University of Texas at El Paso), Dr. Kristy Daniel (Texas State University), Dr. Yaping Moshier, Dr. Brian Donovan (Biological Sciences Curriculum Studies), Stephanie Daniel, and David Merrill (Thirsties). I'd also like to thank my seventh-grade English teacher, Mrs. Bolthuis, who gave me the best piece of writing advice I've ever received that I pass on to my students: just write; worry about editing later. Without this advice, neither this book nor any of my other works would have been written. To my teaching mentor, Dr. Jacalyn Newman, thank you for encouraging me to teach biology and especially to teach it creatively. Thank you to Frances Rabon (The Doula Mommy) for the headshot found at the end of this book. To my parents, thank you for instilling the love of science and creativity in me. Finally, I'd like to thank my copyeditor, Shelley Widhalm, and book designer, Colin Graham, for their help turning my thoughts and ideas into a marketable product.

FOREWORD

I have been an educator for more than 20 years and still feel passionate about helping others learn. Perhaps amusingly, I didn't really know I had that passion until after I already had a job as a biology professor. I knew I loved science of all sorts, and especially anything about the brain; accordingly, my undergraduate degree and PhD are both in neuroscience. I made it all the way through my PhD program and several years of my postdoctoral appointment before I realized that working in a lab, no matter how much I enjoyed thinking about science, didn't quite work for me. I wanted to spend time talking to others about lots of different topics in biology rather than spending long hours looking through a microscope or running gels. So, I made the most logical shift I could think of, which was teaching. I had almost zero preparation, aside from being a teaching assistant for one semester in graduate school, which, although beginning to change, is still typical for most scientists. Thus, it was quite challenging for me to figure out how to educate others, even on subjects I found fascinating, because I'd never really thought about how to communicate to students who didn't have as much preparation as I did. Luckily, it also immediately felt exciting and worthwhile. Soon, I became interested in why students struggled so much in learning biology, and I began a different kind of research career that has involved merging my love of science and my belief that we need to make fundamental changes in the way we engage students.

Early on, I taught genetics to non-STEM majors. This was the most rewarding and yet also most challenging of my teaching assignments because the students, despite their interest in human inheritance, were almost universally unable to understand information in the textbook. It was a mystery to them, the way figures were drawn, the way language was used, the way facts tied together into a story. Even now, teaching genetics to biology majors, the problem remains. Thus, my research focuses on understanding the barriers and struggles students experience while learning, and I work every summer with college faculty around the country to help them improve their teaching practices to better engage students. Ultimately, we scientists need to figure out how to communicate better with non-scientists.

Biology Everywhere: How the Science of Life Matters to Everyday Life is exactly the sort of book that can speak to non-scientists. It not a textbook, yet it describes many of the topics often taught in an introductory biology course for either undergraduate non-science majors or in high school or college general biology classes, plus it covers a few unique subjects not typically found in formal biology coursework. In that sense, it is the perfect book for anyone who says they are "not a science person." This depressing phrase often comes up when interacting with non-scientists—people perceive they won't be able to understand biology, or that it won't be interesting, and then give up on their learning process. And yet, if we can draw them in just a little to the joys of biology, they often want to embrace the challenge. It is critical that everyone understands the topics presented in this book, as they will impact decisions we make as citizens about ethics, climate, and medical care, to name just a few. This book can help people understand that science is simply part of our human experience.

Dr. Melanie Peffer has long been passionate about communicating with non-scientists by connecting biology concepts to their daily experiences. Her training and career have merged biology and education, as she has drawn from research in cognitive science and the learning sciences to hone how she helps others learn. Her unique voice and perspective allow her to connect with readers to show them that biology is quite the opposite of difficult to understand. She has filled the following chapters with relatable stories that help explain the fundamentals of biology, along with personal and funny anecdotes that make the concepts easy to remember. Packed with information, *Biology Everywhere* takes questions and scenarios we encounter in daily life and turns them into the fodder for biological explanations, never failing to draw the reader in. Although it is not a textbook, I can imagine using the book in this way, because anyone can read and learn from this book, from young adults to experts in biology.

I hope you enjoy the journey!

Jennifer Knight
Boulder, CO

Associate Professor, Molecular, Cellular, and Developmental Biology,
University of Colorado
President, Society for Advancement of Biology Education Research,
2018-2019

February 2020

CHAPTER 1

WHY BIOLOGY EVERYWHERE?

WHEN I WAS growing up, anytime we went outside the house, it would turn into a science lesson. My dad was always pointing out and explaining various things. I remember being in elementary school flying somewhere and my dad explaining to me why the flaps were moving in a particular way when we were coming in for a landing and why the wheels go up and come down during the flight. When aerodynamics was explained in terms of what occurred in my immediate environment, they didn't seem so intimidating. Every year during our family trip to the beach, I hear from my dad about the galvanic corrosion occurring between the pipes and the pipe holders hanging above the porch area, each made of a different type of metal. Although I'm not sure my dad, a metallurgical engineer, ever fully recovered from my decisions to study "things that crawl and stink" (biology) rather than follow in his footsteps in engineering, I've adopted a similar mindset about the biology all around us and now point things out to my husband, son, and students.

I've been privileged over my career to work with a wide range of students in very different educational settings. One observation that I've made over and over again, whether I'm working with teenagers, young adults, or senior citizens, is the pervasiveness of their distaste for science. Semester after semester, I had undergraduate students enrolled in my non-STEM majors (STEM is defined as science, technology, engineering, and math) biology class tell me they were no good at science, they hated it, or they were afraid to take my course. This isn't an isolated observation either. Plenty of research supports the distaste and fear people have for biology and the sciences. There is also research suggesting that this distaste and fear of science is likely linked to the rampant science illiteracy in the United States. A quick review of social media reveals several tongue-in-cheek memes about various scientific advances of past decades compared to now in 2020 where it is necessary to state that the earth is in fact, round.

As a society, we are facing critical scientific and biological decisions, and an informed citizenry is important for the continued wellness of our society. Think about the issues you may have heard about: climate change, vaccine debates, the safety of genetically modified organisms (GMOs), and the new genetic technologies that open up Pandora's box of ethical quandaries. My motivation in this book isn't to try to sway you one way or another about any of these issues, but to take the approach that my dad took with me and that I later took with my students and my own child, to expose you to the biology around you in your daily life. My hope is by connecting you with the biology around you that it will help empower you to engage with these issues and make your own informed decisions.

To that end, this book covers much of the content traditionally found in an introductory biology course, but through the lens of how that content connects to our daily lives. The content includes everything from philosophical perspectives of what science is and isn't (Chapter 2) to how we evaluate and make sense out of science information (Chapter 11). Also, unlike regular biology courses, I added several chapters on special interest topics that would not be traditionally covered either due to time constraints or the interdisciplinary nature of the topics, or both. For example, in Chapter 12 we'll examine the art of biology and the biology of the arts.

Biology Everywhere opens with a description of what science really looks like and how it differs from other disciplines of inquiry, such as religion or philosophy. This chapter is probably unlike anything you have seen in biology and other science classes and may surprise you. Following our introduction to what science is and is not, we then turn to chemistry (Chapter 3). Why is chemistry mentioned in a biology book? Well, chemistry is foundational to biology (as is art, but I'll come back to that in a minute). Chemistry explains how we convert what we eat into energy, and as I'll go into in the chapter, we are essentially big chemical reactions walking around. All of our experiences with the world come back to ongoing chemical reactions in our bodies. From chemistry, we'll move into other molecular topics, including cell biology and cancer (Chapter 4), why (most) of life on earth depends on plants (Chapter 5), and basic genetics (Chapter 6). Following the genetics chapter, we'll go into genetic technologies and ethical, moral, and safety considerations associated with these exciting new technologies.

Following our discussion of molecular topics, we'll spend two chapters discussing biology on a larger scale, namely on evolution and the diversity of life (Chapter 8) and ecology and the interconnectedness of life (Chapter 9). Chapter 9 will include discussion of conservation and what it means to go "green" and if so-called "green" options are actually better for the environment. I've also included some interdisciplinary topics that are not typically taught in biology classrooms but that I thought may be of broad interest, including the psychology and biology of child development (Chapter 10), how we reason about biological decisions (Chapter 11), the arts and biology (Chapter 12), and the business of biology (Chapter 13).

Two of these chapters specifically cross boundaries between biology and psychology. Chapter 10 spans courses traditionally taught in biology departments (developmental biology) and psychology departments (human growth and development) to give a well-rounded view of the process of going from two separate cells, one from each of your biological parents, to becoming a fully functioning adult human. Chapter 11 also draws heavily from psychology but is unlike the other chapters in this book. Instead of describing biological phenomena as it relates to our daily lives, it discusses cognitive phenomena that explain how we engage with and reason about our biological decisions in our daily lives.

Chapters 12 and 13 take two fields that are traditionally thought of as separate from biology but are actually quite interrelated. Chapter 12 examines art and biology. This includes a discussion of art as a foundational practice for advancing biological research and what biology and neuroscience research can tell us about how we as humans interact with the arts, particularly music. Chapter 13 examines the business of biology, how biological research works (particularly research on humans), how it is funded, and how clinical trials work. We close with perspectives on the future of biology education and the next steps we can take from what we learned in this book (Chapter 14).

Biology Everywhere was inspired by my experiences teaching non-STEM majors who took my class based on a university requirement for all students to complete a certain number of science credit hours. My former non-STEM majors biology students hailed from many disciplines including art, business, psychology, and education. My conversations and pedagogical approach while working with these students were a major inspiration for this book. I found that if I approached biology content from the perspective of how the issues we were discussing in class related to their lives, personal interests, and majors that students changed their minds and attitudes toward science. I'm honored by the multitude of students who left me emails, notes, and messages scribbled onto their homework at the end of the semester that my class was their first ever positive experience in their science education. My hope here is to expand that impact beyond the classroom to a wider audience who wants to know how biology relates to our everyday lives and about the big issues we are facing in society today.

Although I'm sure people who love biology or biology education will enjoy this book, it is specifically written for people who aren't confident about their ability to engage with biology or science, who want to see how biology relates to their daily lives, and who want to be able to make empowered, evidence-based decisions about the multitude of biological questions we find ourselves facing every day. My goal is to reach those of you who have had prior poor experiences in biology or science classrooms and foster confidence in you to be able to engage with the biology that surrounds you. I've done my best to cover a wide range of biological topics to give an overview of the breadth of biology, while also presenting opposing sides of several issues in today's society. This work is not intended to be textbook (although it may be a useful text in certain classrooms), nor an exhaustive description of any one area, but rather an overview of our rich and fascinating biological world through the lens of our everyday experiences.

The beginning is a good place to start, so let's turn to the first content chapter and discuss what science is (and isn't).

Chapter 2

…BECAUSE SCIENCE

I FEEL LIKE every time I open social media I'm inundated with headlines that tell me "you should do this-that-or-the-other because SCIENCE" or "SCIENCE says that …"

Somehow in our common vernacular, science has become a noun rather than a verb. We see this with science terms as well. My favorite being the use of the word *theory*—we use it over the course of our days to describe a hunch or an intuition about something. However, to a scientist a theory is anything but a hunch—it is a well-supported explanation for a phenomenon. When I say well-supported, I am referring to evidence. Evidence is key when trying to figure out how to define science. What is science? Science is a way of exploring the world around us, of generating new insights (or evidence) that offer some explanations as to how, when, where, and what processes underlie the phenomena that exist in our world.

Why does it matter if journalists wish to describe SCIENCE as some all-knowing indeterminate noun? Because it fuels science illiteracy. It perpetuates the idea that science is an authority figure that must be believed at all costs. It suggests that once science says it's so, then it stays that way forever. It fuels pseudoscience by presenting poor evidence as truth if we frame it in the light of "well, science says so." We'll come back to this idea of the source of science knowledge in our discussion of evidence evaluation in Chapter 10.

Misuse of science and terms like "theory" becomes even more problematic when you hear "well, it's only just a THEORY," because again we are seeing conflation between what theory means to a scientist versus a non-scientist. Did you wash your hands after you used the toilet? Well, you must ascribe to germ theory. Germ theory explains how microorganisms such as the *E. Coli* in your poo can make you sick if you ingest them. It isn't a hunch that someone had one random day to encourage people to buy hand soap and hand sanitizer. Multitudes of scientific evidence support that washing hands after using the toilet minimizes the spread of dangerous disease.

Washing our hands and moving on from the toilet, let's get back to science and, even more importantly, the subject of this book: biology. Biology is the scientific study of life. To understand biology, you must first understand the scientific principles that make up any science discipline. After all, introductory textbooks very often start with the oft maligned and typically skipped WHAT IS THE SCIENTIFIC METHOD? chapter.

It's the first day of class, and you sit down and the teacher says, "Today, I'm going to tell you about the scientific method. First you make an observation. Then a hypothesis. Then you do a test. Then you make a conclusion. The end." Even if perhaps you've heard some variations on this theme, all too often science is presented as this neat orderly checklist of things you do until you get results. A commonly used analogy is that the scientific method is like following a recipe. You follow the steps in the recipe and voila! Dinner is served! Even when following a recipe, things go wrong in the kitchen. Also, like science, there can be more than one acceptable way to cook a meal, and certain methods of cooking don't fit all types of food. For example, you wouldn't necessarily make a Thanksgiving turkey in the microwave, especially if you had an oven or crock-pot at your disposal. We'll come back to this idea of the method fitting the question momentarily.

I operated under the wrong conception that there was a neat and tidy scientific method until I started working on my doctoral degree. I had a rude awakening in my first year of graduate school about how the act of doing science often goes horribly wrong or nowhere at all. If I was a good girl and followed all of the steps, I'd publish papers, earn some grant funding, and get my PhD, yay! Part of becoming a scientist involved rejecting that notion and the realization that success, in science or life, involves facing an awful lot of failures, encountering dead ends, trying new things, and tinkering to find something that works. The harsh realization for me was that I didn't figure this out until I was a PhD student. I made it through AP biology in high school and earned a bachelor's degree in biology ... but I didn't figure out this critically important piece. After many conversations with students and adults outside of STEM (science, technology, engineering, and math) fields, maybe it is not surprising how many people view science as boring, hard, and abstract, following the belief that once we know something (SCIENCE says it's true!), we know it forever. If we followed the scientific method to get there, it must, therefore, be true.

It also may not be a surprise to realize that science literacy rates are low. Historically, students were taught the wrong thing about science from elementary school through higher education. The "simple inquiry" model has traditionally prevailed. Furthermore, science is presented as so inaccessible and abstract that people develop negative emotions toward it and no longer wish to engage. Introductory biology courses have a particularly bad reputation since research has indicated that students learn more vocabulary in a first semester biology course than in a similar first semester language course. This sends the message that science is straight forward and all about memorizing facts ... which couldn't be further from the truth in practice.

Are there efforts to change this? Absolutely—science practices, such as inquiry and argumentation, are emphasized in the Next Generation Science Standards. The Next Generation Science Standards debuted in 2013 to improve K-12 science education. Great! What about everyone else? What about all of the adults out there who have the wrong idea about science or don't feel confident in their ability to engage with science? Movies and other media portrayals certainly don't help science's profile. How many movies, books, and TV shows have been produced where the scientist is the bad guy? Hey, I'm a scientist over here; when do I get to be the good guy?

As I have said to my burned-out by science, terrified non-STEM majors many times, I apologize to you on behalf of science educators everywhere. I am sorry you had such a negative experience with science at some point in your life, whether your chemistry experience was solely about memorizing the periodic table or your biology experience was being grossed out while dissecting a cat ... and wondering the whole time where that poor kitty came from in the first place. My hope is that on the next few pages I can give you an eye-opening, interesting, and accessible discussion as to what precisely science is really all about.

THERE IS NO UNIVERSAL SCIENTIFIC METHOD.

Consider how you transport yourself to work or school every day. Do you drive? Take a bus or train? How about a car or van pool? Maybe you walk or ride your bike. Maybe you work from home and "going to work" means walking across your residence to your desk. For the sake of this analogy, let's say you drive. What do you drive? Is it a car, truck, van, or something else? Old or new? Manual or automatic transmission? Blue, red, or black? Like the "scientific method," commuting sounds fairly straightforward—it is the way we get to work or school. However, when we dig a little deeper, we realize there are many different ways to commute, all valid—and when we apply this to the scientific method, we realize that there are just as many ways to do science. Going back to our cooking analogy, what makes the method appropriate has more to do with the question (or goal) of the research, hence the idea of cooking a turkey in the microwave versus the oven or crock pot.

One of my favorite examples of the various acceptable scientific methods was after the New Horizons space probe flew past Pluto in 2015. During the flyby, scientists at the National Aeronautics and Space Agency (NASA) were able to collect reams of data. A friend of mind commented, "I wonder how they can publish anything since they don't have a control group." Hmmm—many of us can reflect on participating in school science fairs and remember needing a control group, or an untreated group, to get full marks. Planetary science, along with sciences such as paleontology, is observational and theory-building or inductive in nature. Observations are made, data is collected, and inferences are drawn to generate explanations for certain phenomenon. It isn't appropriate or necessary to have a control group.

Planetary scientists who are studying the data generated by the New Horizons probe are still scientists and still using a scientific method, even without a control group. In other disciplines, control groups (often plural) are the gold standard of doing good science. In molecular biology, the discipline I trained in, it is not uncommon to see many different types of controls. There is a group where you don't do anything, or an untreated control. There could also be a vehicle control. The vehicle refers to whatever you used to dissolve your compound of interest. For example, if you are making hummingbird food, water is the vehicle. Then you have your experimental condition, which has your molecule of interest, plus whatever you dissolve it in. In the case of our hummingbird food analogy, it would be the water and the sugar. This type of work is called deductive or theory-testing and is the type of science with which most of us have familiarity.

What is universal to science? What differentiates science from pseudoscience? It boils down to evidence. Not a hunch or solely an anecdotal observation but empirically collected evidence. Even more importantly, high quality evidence. Evidence can be in many forms but is typically thought of as collected and analyzed data. For example, when I was a student, I was interested in how male or female cells expressed different genes after treatment with a drug commonly used in women at risk for preterm labor. If I only do the experiment one time, it would not be considered high quality evidence. Maybe I do something wrong and my observation is a fluke. If I can repeat the experiment over and over and get the same result, and even more importantly someone somewhere else can repeat what I did over and over and get the same result, this strengthens my evidence and consequent conclusions. To control for possible bias (I'll talk more about that in a few pages and in more depth in Chapter 10), I could treat my cells blinded, where I don't know which treatment is which (but someone who is not working on the project does) until the results are in. This adds even more strength to my evidence and is also why double-blind randomized control trials (which we'll cover in Chapter 13) are the gold standard for evidence-based medicine. Neither the patients participating in the study nor the physician know who is in what group—only the external researcher.

Are anecdotal observations important? Of course—but the evidence provided does not carry the same weight as an empirical study. For example, if a doctor sees a patient with an unusual inexplicable condition, they may publish what is called a case study. A case study includes detailed information about their observations and may speculate at the root cause of the disease. A case study may go on to inform a larger study at a later date. This is also a nice example of my next topic, which focuses on how science is done by groups of people.

SCIENCE IS DONE BY GROUPS OF PEOPLE.

Close your eyes and think of what a scientist looks like when they are doing science. Do you imagine the person working alone in a lab setting? Or maybe talking to one or two other people? Did you think of one person standing up in front of a large audience discussing results? Or video conferencing with someone on the other side of the world? Or what about someone reading the works of a long dead scientist to inform their current work? Or reviewing the results generated by another group of scientists under consideration for publication in a reputable journal?

There are many different processes to generate scientific information, and there also are multitudes of ways scientists collaborate with one another. An interesting artifact of this process is the difficulty in getting completely new information published because of the expectation that your ideas and work build off of the ideas and work of others. Really novel ideas aren't always widely accepted at first. Scientists in all disciplines work together to solve problems, share ideas, and review each other's work. Scientists in a certain area will come together yearly at conferences to discuss and exchange ideas. It is not uncommon for research projects to be international efforts. Ideas are communicated across time as well—published works are a way for scientists to share their findings and ideas long after they have died.

In order for research to be accepted as credible, or for funds to be received to do research, the research must go through a peer-review process. In the case of writing a grant, scientists must give exhaustive descriptions of what they plan to do, why they are doing it that way, and the areas where things could potentially go wrong and what they plan to do about that. A typical proposal going to either the National Science Foundation or the National Institutes of Health is 15 single-spaced, 10-point font pages, alongside a wide variety of ancillary documents that support why you are the best person to do the work and why the place where you plan to do the work is the best. These proposals get reviewed by other scientists in the field, and if you have a great idea and can convince your peers (plus the governmental officials who make the final funding decisions), you get funds to do the work.

After the work is done and you write it up for publication, potentially passing through another stage of peer review if you are proposing to use either animals or humans as part of your research, it goes through even more peer review. The final work can be reviewed two, three, or even more times before it is finally deemed acceptable for publication. Each time the work is reviewed, the reviewers send questions and required improvements that must be addressed before moving on to subsequent stages of review. If reviewers find a fundamental flaw in the research, it may never be considered high enough quality to be published.

Curious about the history of a scientific paper? How many times it was reviewed is often included in the manuscript. For example, if you look at one of my papers from my graduate work, shown in Figure 1, you can see the history of my article, including that it was returned for one round of modifications and additional experiments before it was accepted for publication.

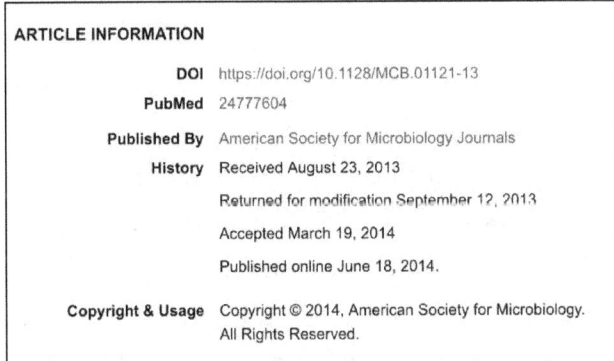

Figure 1. A Screenshot from my first research publication documenting the peer review process.

Other important information in the paper includes the publisher—a high quality well-known publisher is likely to have a better peer review process as well. As we'll come back to in Chapter 10, this evidence of an external peer review and subsequent revisions is one way of determining the quality of a research article.

SCIENCE IS BY NO MEANS AN UNBIASED PROCESS.

Although the peer-review process attempts to assess evidence and findings based on their own merit and is often blinded to prevent bias, bias has a tendency to sneak up anyway. One of the pitfalls of using scientific evidence is the fact that how we use and interpret evidence is subject to our own cultural beliefs and biases, a subject we'll return to in Chapter 11. It is very difficult, if not impossible, to fully divorce our biases from our work. Although we put forward the ideal that by adhering to a scientific method, we remove all biases—because of how our brains work, this is not fully possible.

We can attempt to mitigate bias by blinding experimental conditions or requiring a double-blind review (where neither the reviewer nor the study author knows the other's identity), but bias can still sneak in. For example, let's say you've made a very interesting observation. You've repeated the experiment multiple times and saw the same result. You were careful and blinded everything. Someone else on another floor of your building was able to repeat the result. Awesome, you managed to generate some good evidence. Now for the hard part—what does the evidence mean? How you interpret that evidence is going to be shaped by many things—your background knowledge, your initial hypothesis, the conversation you recently had with a colleague, what your supervisor thinks it should be, even things like your religious and cultural beliefs.

One of my favorite examples of how culture influences science was during the development of the first birth control pills. The scientists had three candidate molecules for testing in humans, but one molecule had to be eliminated. The molecule was eliminated from consideration not for any scientific reasoning, but because of the religious beliefs of the owners of the company that produced it. The owners refused to sell it to the scientists because of its intended use as a contraceptive. There was no scientific reason for its elimination, only a cultural one.

Another place I see the influence of bias on interpretation and use of scientific evidence is in so-called "evidence-based" online parent support groups. Scientific evidence can and does change over time in light of new findings or new technologies that allow us to collect data in new ways. It is common for a practice to be evidence-based (or not evidence-based), but in light of new evidence, recommendations around the usage or implementation of a practice will change. In terms of consumer products, initial testing may suggest the product is safe to use, but new evidence may overrule the current recommendations. New interventions may lack evidence initially while a substantial body of work is generated to support their implementation. There is a difference between evidence-based today, and what will be evidence-based tomorrow. Therefore, it can be easy to find "evidence" that supports our biases, "something we'll return to."

Since life tends to be murky and often lacks hard and fast answers, it is easy to find data to support whatever you want and to discount the other stance. Let's say, for example, you have decided that any use of essential oils is bogus. It is easy to say, "See! There is no evidence to support this!" Just because there isn't any evidence does not necessarily mean you can support a negative. No evidence is just that, no evidence. It does not mean the opposite is true. On the other hand, if you think there is scientific evidence supporting using essential oils, you might look in the literature to find that certain compounds found in particular oils do have some evidentiary support, be it only in cell lines and not humans. Of course, the companies selling essential oils typically have non-peer reviewed studies funded by the essential oil companies (which is a conflict of interest since the companies are biased toward wanting to sell you essential oils), and, of course, those papers claim that essential oils will treat XYZ.

This gets even more heated with issues that have good evidence on both sides. Take infant feeding for example. There is a huge amount of evidence supporting breastfeeding, with new research on the beneficial biological properties of human breast milk released all the time. However, there also is evidence suggesting that the benefits from breastfeeding may be due to other factors like socioeconomic status. Uneducated breastfeeding can pose health risks to both infants and mothers. For example, without proper education on how to ensure that the baby is getting enough sustenance at the breast, terrible situations can arise where the baby's health is in serious jeopardy from dehydration or lack of nutrition. The immense pressures felt by women to breastfeed and a general lack of support (at least in the United States) also can cause significant mental health concerns for the mother. Lack of proper education on how to manage engorgement can lead to the mother developing mastitis (milk infections), which can be serious enough to warrant a hospital stay. So, the internet is rife with circular arguments, all with good evidence, stating that "breast is best," "fed is best," and finally, my personal favorite, "informed is best."

The position taken is almost always a factor of your cultural and family dynamics. If you and everyone you know was formula fed and went on to be healthy adults, it is very easy to fall into a "fed is best" category. If you are a lactation consultant and know the research on the truly awe-inspiring properties of human breast milk, it is easier to fall into a "breast is best" mentality. If you're like me and many other mothers I know, you know breastfeeding isn't a cake walk and that it can be very undermining to be told "fed is best." Knowing that it's not uncommon to need formula or donor milk, or knowing how very, very sick you can get from mastitis, it is easy to align with the "informed is best" camp. No matter what camp you fall into, you are guaranteed to find evidence to support it—because if you are human, you have biases.

CREATIVITY BELONGS TO SCIENCE AS WELL AS TO THE ARTS, AND THE BEST SCIENTISTS ARE THE CREATIVE ONES, NOT THE ONES WHO ARE ALWAYS RIGHT.

Whenever I go over how science works in the real world, this one always takes my students by surprise. Although we hear about STEAM (adding in the A for arts with science, technology, engineering, and math (STEM); more on this in Chapter 12), there still tends to be a disconnect with realizing what the arts have to do with science. At a fundamental level, good science requires the same amount of creativity and thinking outside the box that is used by songwriters, musicians, writers, and artists. Many of my past non-STEM students were amazed when they began to draw parallels between the creative processes necessary to science and those in their own arts majors.

Where do we see creativity in science and biology? In looking over past results and finding new interpretations. In generating new methods or new solutions for problems. In the final interpretation of results. A great example of creativity was when Osamu Shimomura, a Japanese biologist, isolated the protein from jellyfish that fluoresces, or gives off, green light. Scientists eventually realized that you could take green fluorescent protein, or GFP, and attach it to other proteins of interest (which are too small to be seen individually). Then you can see where the light is and indirectly identify the location of the protein itself. Think of it like putting lights on a child's Halloween costume to keep them safe trick-or-treating—a driver may not see your child (like you can't see an individual protein if you look down a microscope), but hooking up a light makes the child (and the protein) visible. It seems simple now, but it was a creative leap to go from jellyfish giving off light to revolutionizing molecular biology—one that was eventually recognized by a Nobel Prize.

It is easy to think that the most successful scientists are the smartest ones who were right more often than not. The scientists we remember are the ones with the most creative ideas, the ones that rethink the obvious or connect the dots in a new way. One of my first lessons about scientific research came from my dad who told me that the secret to a career as a successful researcher was to make the obvious connections that no one has seen yet. Another early lesson during my training as a scientist is that the process of science is super messy and involves failing ... a lot. It involves having what my graduate mentor termed "crazy ideas" that are creative but can and often do fail. While I was in graduate school, many of my mentor's crazy ideas ended up not going anywhere, but one idea that serendipitously sprouted from another student's project (and led to me sending a late night email after looking at the results saying, "This was odd—did you expect this to happen?") eventually resulted in an entire first-author journal publication for me. It really is not that different from an artist having an idea and trying something new—it might totally flop, or it could be the next big thing.

KNOWLEDGE GENERATED BY SCIENTISTS CHANGES IN LIGHT OF NEW EVIDENCE, AND THEREFORE IT IS NOT POSSIBLE TO PROVE ANYTHING DEFINITIVELY.

Galileo ran afoul of the Catholic church for suggesting a heliocentric (the earth travels around the sun) rather than a geocentric (the sun and planets travel around the earth) model. The geocentric model certainly makes sense if your evidence is limited to what you can see yourself without any equipment. Over the course of the day, you can see sunlight move. My house faces east, so in the morning my eastward-facing bedroom is bathed in light, whereas the westward bedrooms only receive light in the afternoon as the sun moves westward. Peering at the night sky, a keen observer can pick out the planet Venus as it traverses the sky (planet after all does mean "wandering star"). This evidence would all suggest that heavenly bodies are traveling around the earth ... unless you possess a telescope. Galileo's observations of Jupiter's moons orbiting not around Earth, but Jupiter, in addition to his observing the changing phases of Venus, all supported the fact that not everything revolved around the earth but that the planets actually revolved around the sun.

Galileo's observations with a telescope illustrated how new evidence can be generated alongside a technological advance. He could see something that was previously unobservable. We now can do increasingly sophisticated studies of gene expression as molecular technologies like gene sequencing become both cheaper and more robust, allowing us to make observations and generate new evidence that was previously inaccessible. The rise of the field of epigenetics, or how physical "marks" on our DNA can influence gene expression, is drastically changing how we understand everything from intergenerational trauma to cancer. Thalidomide was touted as a wonder drug for curing pregnancy-related nausea until evidence came to light demonstrating that it caused birth defects. The new evidence (birth defects) changed its usage in medicine (don't use it in pregnant women). The reason I posit that science can't prove anything is because what we know changes—and what is the best thing one day may not be the next if there is new evidence. That is the beauty of science knowledge—it is a constantly evolving entity based on what we know changing over time in light of new evidence.

Since scientific evidence always is changing, and there is new evidence coming to light, it leads to another critical point best summed up by Stuart Firestein, a neuroscientist at Columbia University. He said science is best described as "thoroughly conscious ignorance"—the more we think we know, the more we realize we don't know. In medicine, a good example of this is why people respond certain ways to treatments—just because we don't know something today doesn't mean the evidence won't exist in the future to describe why person A responds one way to a treatment and person B the opposite. The relationship of evidence to types of knowledge brings us to our next key point—the relationship among hypotheses, theories, and laws.

Hypotheses become theories, and theories become laws, right?

In every non-STEM majors biology class I've taught, one of the most common misconceptions students have on the first day of class is that hypotheses become theories, and theories become laws. This also may be one of the drivers of the dismissive "well, it's just a theory" statement. Hypotheses and theories propose explanations with theories being much broader explanations. I like to think of hypotheses as answers to a question, whereas a theory has broad applications on a much larger scale. For example, I might hypothesize that a particular cell has a certain function, but cell theory explains why we have cells and where they come from in the first place. Laws are typically describing some kind of physical relationship that is unchanging. For example, Maxwell's equations are used to describe electromagnetism—it is a numerical way of describing a physical relationship. The equation remains consistent, because the relationship is the same. Hypotheses can underlie either a theory or a law, but a theory would never become a law—each is used for describing two very different kinds of scientific information. Theories are broad explanations that are subject to change in light of new evidence, whereas laws describe an unchanging physical relationship that can be described numerically.

Summary

This chapter examined how science and, consequently biology, work in the real world. Science is a process to learn about the world around us, rather than being a collection of facts. There isn't a single process, or method, to explain how science works. Rather, the methodology used is designed to meet the question at hand and can vary widely both between and within different science disciplines. Science is not done by single individuals working alone, but rather in teams, sometimes spanning international or generational borders. Bias plays an important role in how science is done and the results are interpreted. Perhaps surprisingly, good science is based on creativity, and creative expression is not constrained to the arts. The power of science knowledge lies in its ability to be revised in light of new evidence, and different types of science knowledge are described in different ways as hypotheses, theories, and laws.

My hope in this chapter is that I have changed your perspective on what science is really all about. The practice of science is far more interesting and dynamic than most people think. Now that we have established a baseline for what science is and isn't, let's zoom in a bit more and discuss chemistry, which serves as the foundation for understanding life.

Chapter 3

Chemistry, Essential Oils, and Venom

Before you flip back and look at the front cover of the book (isn't this a book about biology?), you'll see that yes, you read the title heading correctly, this is a chapter about chemistry.

We started in Chapter 2 by talking about what science (and therefore biology) is really all about and now we move into chemistry. You may ask, why spend an entire chapter of a book about biology on chemistry? Whenever I teach non-STEM majors biology to undergraduates, I use the analogy that the chemistry chapter is the musical equivalent of playing scales; it's foundational for understanding the rest. My general chemistry professor in college started out the two-semester general chemistry course by talking about chemistry as a central science. You can learn chemistry for chemistry's sake, but it also connects with and is fundamental to physics and biology as well.

What does chemistry have to do with biology after all? How does chemistry relate to my life? At the most fundamental level, you are chemistry. All you are is a walking, talking, breathing bag of aqueous (water-based) chemical reactions. How are you reading this text right now? A series of chemical reactions triggered by light hitting your eyes. How did you turn the last page? A series of chemical reactions triggered by your brain. Those chips that you are munching on while reading? They get processed into energy and waste through ... you guessed it, chemical reactions. Chemistry is all around us—it explains why fireworks are a particular color, why gasoline is better as fuel for your car than water, and how soap cleans dirty hands. Even if you may cringe when you first read the chapter heading of "chemistry" and immediately think I'm going to start reciting the elements of the periodic table, the chemistry of our lives is more prevalent and interesting than many realize.

If you had to memorize the periodic table or ion charges ($Ca^{2}+$ anyone?), like many of my former students, and are dreading reading this chapter any further, don't worry, we aren't going there. Instead, let's talk about chemicals. Chemicals are bad, right? It doesn't take long over the course of our day-to-day lives to hear an example of how chemicals are *bad*. They make us sick, can give us cancer, and they are dangerous. I recently saw someone post in a mom's support group I'm in that she uses essential oils to clean her house because she *didn't want chemicals* in her house.

So, what is a chemical?

Let's take a step back—what is a chemical? According to the Cambridge dictionary, the word "chemical" can be used as either a noun or adjective. As a noun, or thing, it is a "basic substance that is used in or produced by a reaction involving changes to atoms or molecules." As an adjective, it is "of, involved with, relating to, or made by using chemicals or chemistry." So what does the dictionary mean by saying it is a basic substance? Let's consider the chemical you are probably most familiar with that is also arguably the most important chemical for understanding life: dihydrogen monoxide.

Ah, dihydrogen monoxide or hydric acid. Too much or too little will definitely kill you. It can burn you or cause your car to rust out. One hundred percent of people who died of cancer had dihydrogen monoxide present in their bodies at the time of death. It causes millions of dollars in property damage annually, and if you live in a dry climate like me, your monthly dihydrogen monoxide bill can cause high blood pressure and aggravation.

Heard that one before? It is a common internet joke designed to make fun of the lack of science literacy that is prevalent in society. Rather than continue to string you along, let's break it down. Di – di means two. Hydrogen (H) is an element, or the smallest "thing" that can be broken down and still be that same "thing." Mono – mono means one and oxygen (O) is another element. Therefore, dihydrogen monoxide is something with two hydrogens and one oxygen. Also known as H_2O ... or our good friend, water. Yes, before you spit out the water you just drank all over your book, it is indeed a chemical, and you absolutely need it to sustain life.

WATER IS A CHEMICAL AND ESSENTIAL FOR LIFE.
Life as we know it requires water. This is why the search for extraterrestrial life revolves around the search for water. Water has several interesting properties that make it conducive to supporting life. For example, let's consider an oak tree. How does an oak tree get water from the ground to its leaves? Unlike an animal, a tree does not have an integrated pump to move liquids—but it still

needs to transport liquids and nutrients. The reason we can have a tall oak tree is because of the unique properties of water. Water is sticky—it sticks to itself and to anything else through hydrogen bonds. Think of it like water molecules holding hands. The ability for water to stick to itself (cohesion) and to stick to other things (adhesion) explains why an oak tree can get water up to the tops of its highest branches. You also can see this phenomenon in action in real time when you stick a straw into a cup. What does your beverage do? It "climbs" up the straw, because it sticks to both itself and the straw, allowing it to "move" up the straw. Water molecules that are evaporating (or turning into gas) at the end of the straw (or through pores in the leaf of a tree) help to pull the chain of water molecules up to its top.

What else can water do? It is a great insulator. I grew up on Lake Michigan and it would stay warm later in the year because the water would hold onto the heat. Anyone who lives near or on a large body of water knows that it stays warmer and cooler longer when the seasons change because water resists temperature changes. Remember when I mentioned earlier that we are walking bags of water with chemical reactions going on? Water helps insulate us, too. One of the reasons we sweat when we're hot is to release heat—as the sweat evaporates, it takes excess heat with it. Water also is unique in that it expands when frozen. Why does this matter? Well, this is why you don't want to leave a sealed beverage in the freezer too long and why animals can live in ponds over the winter. When water freezes and expands, it also becomes less dense and will therefore float. In the case of a pond, the frozen water on top floats and creates nice insulation for the critters living below the ice.

Natural and Synthetic Chemicals.

So water as a chemical is okay then, right? It is a *natural* chemical after all. As a society, we are beginning to scrutinize our foods and consumer products in an effort to avoid dangerous chemicals or toxins. A growing movement suggests that natural should be sought at all costs. This is particularly problematic when lifesaving modalities such as vaccines are disparaged as "unnatural" (even though the biology that explains how they work is based on a natural process) or full of "dangerous chemicals." Let's step back for a second and remember that a chemical is a "basic substance that is used in or produced by a reaction involving changes to atoms or molecules." Chemicals include things that are perfectly natural but could sound intimidating if listed by the chemical name—such as dihydrogen monoxide, sodium chloride (table salt), and 3,7-Dimethyl-1,6-octadien-3-ol. The chemical 3,7-Dimethyl-1,6-octadien-3-ol sounds really gnarly, but it is linalool—a plant-produced alcohol. You probably know exactly what it smells like, too—it's the main chemical given off by the *Lavendula* plant, and it is the scent we call lavender. It's also the main component of lavender essential oil. And believe it or not, in addition to being an alternative treatment for ages to treat anxiety and depression, linalool, as demonstrated by recent cell culture studies,[1] acts through the same cellular pathways as Selective Serotonin Reuptake Inhibitors (SSRIs), which are the frontline medicines for treatment of anxiety and depression. It also impacts the activity of the N-methyl-D-Aspartate (NMDA) receptor, which is targeted by alcohol and other drugs of

---

[1] López, V., Nielsen, B., Solas, M., Ramírez, M. J., & Jäger, A. K. (2017). Exploring pharmacological mechanisms of lavender (Lavandula angustifolia) essential oil on central nervous system targets. Frontiers in pharmacology, 8, 280.

abuse. What is particularly interesting is that ketamine, a drug used during anesthesia, influences NMDA activity, and was recently approved in the United States for treatment of treatment-resistant major depressive disorder.[2] Does this mean we should eschew modern psychiatry and pharmacology? No, because knowing cellular targets isn't the same thing as understanding how a disease process works in the entire body. Cells do not necessarily behave the same way in a dish as they do within the context of a body. As we will see in the next chapter, cells are constantly in contact with one another throughout the body. Cellular communication, or cell signaling, plays an important role in understanding disease and how different compounds impact cellular function. However, it is intriguing to note the potential parallels from traditional folk remedies to modern medicine.

Consider salicylic acid or taxenes. Both are compounds derived from tree bark but also are used in society in their purified form. Salicyclic acid is also found in willow tree bark, and you may be more familiar with its other name: aspirin. Taxenes are derived from Pacific Yew trees and are used to produce a variety of chemotherapeutics, such as paclitaxel or docetaxel.

Hang on—if you've ever had a loved one take paclitaxel or docetaxel, you know that these medicines can make loved ones very, very sick. Taxenes interrupt cell division (we'll come back to cell division in Chapter 4), so any non-cancerous cells that divide regularly, such as in your hair follicles or gut, also will be affected. This is why people getting treated for cancer are often nauseous and lose their hair. But it's natural! Isn't that better? Not necessarily.

---

[2] Here's information on the clinical trials: ClinicalTrials.gov. Bethesda (MD): National Library of Medicine (US). 2004 Aug 2. Identifier NCT00088699, Rapid Antidepressant Effects of Ketamine in Major Depression; https://clinicaltrials.gov/ct2/show/NCT00088699

Figure 2. Image of cone snail shells. (Photo by Pet/CC BY-SA-3.0)

Some of the nastiest chemicals that exist are completely natural. Take for example venom. All of the venoms I am going to mention occur naturally. Let's start with the cone snail. You are diving in the tropics and see a beautiful shell (Figure 2). It is the shell of a cone snail. It would be a *very* bad idea to pick up that shell. Why? Cone snails hunt by harpooning their prey and injecting venom. If the cone snail feels threatened and decides to harpoon you instead, its venom will cause paralysis and very likely death. Estimates suggest that the venom from a single cone snail can kill 700 people.

How about the Boomslang, a snake native to Africa? This venom is hemotoxic meaning it interferes with your body's ability to clot blood. What is the ultimate result of an untreated bite? You die from bleeding out. The good news is it takes a while to bleed out, which gives much needed time for an anti-venom to work, but it can also lull the victim into a false sense of security thinking that maybe it was a "dry," or non-venom-containing, bite.

Natural isn't necessarily always good. Scientific research can certainly help us develop targeted therapies to mitigate side effects. Pharmacology and drug discovery is becoming very precise, and better synthetic compounds are coming to market. These compounds are not natural but can really help. For example, Gleevac (Imatinib) is a synthetic drug for treating certain types of cancer like chronic myelogenous leukemia by specifically inhibiting a single enzyme called Bcr-Abl.

However, there are certain unnatural substances being found in our foods and consumer products that do warrant our attention. Again, it isn't that the substance is or isn't natural, but it's a matter of what we know about the chemical properties. Have a plastic water bottle handy? Does it say BPA-free? Bisphenol Alcohol (BPA) is used in the production of plastics, many of which we encounter in our daily lives in food packaging. Studies show that BPA can leech into our foods and that BPA can cause health problems. However, the FDA(Food and Drug Administration) says that the levels of BPA in our food are likely subclinical, meaning they shouldn't be at high enough levels to make you sick, but there is ongoing research at the FDA that could lead to policy changes in the future. This is another nice example of showing the process of science. If the FDA uncovers information in future research that there are actually clinically meaningful levels of BPA in our foods, then the recommendation will change. In the meantime, if you don't want to wait to find out, you can seek food or beverages in BPA-free packaging.

THE CHEMISTRY OF EATING.

Where else do we see chemistry in our day-to-day lives? Our diet! "Food" in chemistry terms falls under one of the four macromolecules. Carbohydrates or sugars, protein, fat, and nucleic acids. The first three probably sound very familiar to you, but the fourth tends to trip people up. Nucleic acids store and transmit information across generations and include deoxyribose nucleic acid (DNA) and ribonucleic acid (RNA). Yes, you read that correctly. You eat DNA. All the time. That tomato you ate on your salad last night? Yup. You ate plant DNA. The chicken on your salad? You got a nice serving of DNA with the protein. Perhaps you've heard of various fad diets. Adkins diet? Cuts out carbs and increases consumption of proteins and fats. Ketogenic diet? Eliminates all carbs from your diet inducing a state of ketosis that relies solely on fat and protein to produce energy and is great for short-term weight loss.

Depending on your individual needs, you may need more of one macromolecule than another at any given time. For example, long-distance runners carry stocks of simple carbohydrate "goos" to fuel up quickly during a long run. Simple carbohydrates (found naturally in fruits and milk) get energy to you quickly—which also is why they are discouraged when trying to avoid gaining excess pounds. Complex carbs (like what you find in potatoes) take longer to break down and therefore aren't a "fast" energy source. We'll come back to carbohydrates and weight gain in Chapter 5.

Proteins are interesting macromolecules since they fold up in unique ways. Think of paper origami. If you want to make an origami swan, the paper needs to be folded just so. Proteins are similar; they also need to be folded just so to be able to function. Why does meat change color when you cook it? Because you've unfolded or denatured the proteins. Going back to the origami analogy, you've unfolded the paper and no longer have a swan. Denaturing proteins makes them easier to eat and kills microorganisms present in our food.

What about the fourth group of macromolecules, the fats? Fats are an essential part of our diet. Have you seen labels at restaurants about frying in trans-fat-free oils? What does it mean when you hear people talk about healthy versus unhealthy fats? Isn't fat something that should always be avoided? Fat, particularly healthy fat, is good! You absolutely need it to survive. Our cells (more on cells in Chapter 4) are little bubbles of water and organelles wrapped up in a fat blanket. When thinking about fats, imagine you are packing a box full of sticks. Saturated or unhealthy fat at the biochemical level is like a nice straight stick. It is very easy to pack those nice straight sticks into the box and pile them up. This also is why saturated fats are solid at room temperature, and why they are particularly unhealthy. It's also easy to "pile" these fats up in our bodies. Unsaturated or healthy fats are sticks with kinks. They are really hard to pack into your box. These fats are liquids at room temperature. Finally, we have trans fats, which also are unhealthy. Now, imagine you've tied straight sticks together into small bundles and then packed them in the box. The extra sticks (or extra bonds at the molecular level) mean not only do trans fats pack nicely together, they have more energy stored in them than saturated fat. What happens to excess energy? It appears on your hips, waist, and rear end as fat.

Another fat you may be familiar with? Cholesterol. We need it as part of our diet because it helps make up cell membranes and serves as the starting material for our bodies to produce steroid hormones like testosterone and estrogen. As with anything else we eat, cholesterol is good in low doses but too much leads to health problems like heart disease.

SUMMARY

Let's steal a chemistry term and *synthesize* what we've learned and point out just where chemistry is apparent in our daily lives. Water is the chemical we are most familiar with. Water is important not only for our own health but important for all of life on Earth. The properties of water explain why we sweat, how oak trees grow so tall, and how animals can live at the bottom of a pond all winter long.

We discussed the push toward natural products in society and the good and bad of placing too much emphasis on a natural compound—or conversely, rejecting compounds simply because they are not natural. Using linalool containing essential oils when stressed may seem kooky (and some claims made by essential oil-producing companies are definitely kooky), but there exists biochemical evidence to back it up. Then there are natural compounds like venom which you would never, ever want to ingest no matter how natural. Synthetic compounds can sound scary or be detrimental to your health, but pharmacology and intentional drug design is creating new drugs with fewer side effects. This is in contrast to potentially problematic synthetic compounds that are leeching into our food like BPA, which deserves our attention.

Finally, there is the chemistry of what we eat. Our food falls under four chemical varieties, and a balanced diet contains all four of them. The food and water we ingest fuel the chemical reactions in our body. Where do these chemical reactions occur? Now that we have successfully "eaten our Wheaties" and "mastered our scales," we can move past foundational information and into our first biological topic and one of my favorites: cells.

CHAPTER 4

CELLS GONE WILD: CANCER

Figure 3. My in-laws and I after my graduation from the University of Pittsburgh in 2014. Both my mother-in-law and father-in-law died from cancer within six months of each other. This chapter is in their memory.

WHEN I WAS an undergraduate, we covered cell signaling, or how cells talk to one another, in my sophomore genetics course. That was the lesson that spurred me to declare a molecular biology major. How cells talk to one another is a beautifully orchestrated ballet that occurs on the molecular level. I was fascinated by how all of the molecular pieces fit together in time and space to communicate information.

The name "cell" came about when the first microscopes were developed and a scientist by the name of Robert Hooke was examining a piece of cork. Cork is part of the bark found on a specific type of tree called a cork oak and is used in a variety of practical applications, including wine stoppers. Hooke observed that the cork was made up of many small compartments. These small compartments reminded him of *cellula*, which were the small rooms that monks lived in, and so he named these compartments "cells." Cork, and many other structures, appeared to be made up of small cells. Scientists following in his footsteps would later learn that there are also single-cell organisms.

Our knowledge of cells continues to grow, which is exemplified by our textbooks. For instance, when I took my first cell biology class as an undergraduate, the gold standard text in molecular and cell biology, *Molecular Biology of the Cell,* was in its fifth edition, weighing 6 pounds, 10 ounces (3,014 grams) and containing 1,268 pages, not including the glossary and index. Seven years later when the sixth edition came out, the book weighed 7 pounds (3,182 grams) and contained 1,342 pages, not including the glossary and index. So, in the span of seven years, the amount of knowledge deemed foundational enough to warrant inclusion in the textbook grew by about 6 ounces or 168 grams. That's more than the official weight of a Major League Baseball, which is set at 5 ounces.

Having a textbook that weighs as much as a typical human baby at birth and that is continually growing suggests we know an awful lot about cell biology (and fosters misconceptions that science is nothing more than a collection of facts). If that is the case, why do we still have rogue cells that cause diseases such as cancer? Let's take a look at some basics of cell biology that set up our discussion for our application of cell biology: cancer.

BASICS OF CELL BIOLOGY.

You can take a semester-long graduate level course in cell biology and not make it through *Molecular Biology of the Cell* in its entirety. I know—I've taught graduate-level cell biology using this textbook. I'm sure you remember sitting in a classroom at some point learning about the organelles, which are structures within the cell that perform some kind of function. Although interesting and important, keeping with the spirit of this book, I'm going to keep this chapter focused on cell biology as it relates to cancer, since cancer is a disease we are probably all too familiar with.

What is a cell? It's the fundamental unit of life. You could draw an analogy to an atom, which is the fundamental unit of an element. Cell theory (remember from Chapter 2 that a "theory" is a well-supported broad explanation for a phenomenon, in this case what cells are and where they come from) states that living organisms are made of cell(s), cells are the fundamental unit of life, and cells come from pre-existing cells.

Let's further break down cell theory. Living organisms are made up of cells. Think of cells like Legos. Legos as individual units don't seem like much on their own, but when connected with other pieces create structures that give form and function. Our cells are organized into structures, such as the organs in our bodies, and these tissues give us form and function. For example, the cells that make up your skin tissue help protect your body from infection. Next, cells are the fundamental unit of life. This is as small as you can go and still have a biologically functioning unit. Finally, cells come from pre-existing cells. This is where the old joke comes into play about biology being the only discipline where division and multiplication are the same thing. Cells divide to make new cells. All cells in our body ultimately trace themselves back to the zygote (your first cell, or the egg from your biological mother that was fertilized by your biological father's sperm); in other words, the cells that fused to form a zygote can be traced back to our biological parents.

What are things that cells do to prevent cancer, or rogue cells? Well, for one they talk to one another. Cells signal to other cells over short or long distances. An example of short-distance signaling is with our immune system where two cells come in contact with one another to signal if a cell is "self" or an invader. Long-distance signaling is something we see with our endocrine system. A rush of adrenaline secreted by your adrenal glands primes every cell in your body to respond to a threat. Cells also tell one another to grow, to divide to make new cells, or even to die. One of the ways cells can go rogue and lead to cancer is when the mechanisms of talking to one another are lost. You can draw a parallel between cells in the context of the body and a hive mind. The goal is for the good of the body or hive, not for the individual cell, so most cellular processes focus on the good of the body. A cancerous cell comes about because it is outside of these controls and no longer functions for the greater good of the body.

One example of how cells talk to one another is by telling each other when to divide, a process called mitosis. Sometimes we need more cells. For example, your hair is growing because cells are dividing at the base. This process of cell division is called mitosis. Cell division in the right place at the right time is important for maintaining everything from our skin to our gastrointestinal (GI) tract. We also need to generate new cells in response to a wound. New cells fill in the space where older cells were injured. It is also important for growth. Consider my son for example; when he was born, he was 21 inches long. By his second birthday, he was 36 inches long. Where did those extra 15 inches come from? Cell division.

Sometimes we need cells to die. For example, when we were developing in our biological mothers' wombs, we, at one point, had webbing in between all of our fingers. We don't need this webbing, and so these cells die through a process called apoptosis, or programmed cell death. There is a Star Trek episode from the original series that can be used to exemplify apoptosis. In *A Taste for Armageddon*, the crew of the Enterprise finds itself caught up in a simulated war between two planets. Each planet has simulated attack on the other, but the deaths resulting from the attack are real. The computer tallies how many deaths occurred during the attack and orders people to disintegration chambers, where the "casualties" are then killed. The process is neat and orderly and protects the inhabitants of the planet from the ugliness of war. Apoptosis is also neat and tidy and designed to protect the whole by removing an unwanted part. Keeping cells dividing only when we want them to and killing off un-needed cells, or cells that have gone rogue, are the key ways the body protects itself from cancer.

So much of cell biology is doing things in the right place at the right time. You want cells to be dividing in your hair follicles, or skin, or gut. If you are injured, you want your body to make new cells so you can heal. Same with apoptosis. It is great for clearing out cells that we do not need any more during our development or for getting rid of old and damaged cells. However, uncontrollable apoptosis leads to organismal death. In fact, randomly occurring genetic mutations in genes that control apoptosis usually result in a non-viable fetus.

How do cells know when to divide or die? Through cell signaling. The same way we can look at a traffic light and know when to go or stop, cells receive signals and cues to divide or die. The social nature of cells is critically important to make sure the right cell gets the right message at the right time. It is an amazing orchestration of chatter. Think about mail delivery. Your mail gets to you and only you in a particular timeframe. There is a specific order and process for your mail to get from point A to point B. We've all felt the frustration when this system breaks down and our mail is sent to the wrong address or sorting station. What about in the body? When signals get sent to the wrong place at the wrong time, or when cells become unresponsive to signals, cells can go rogue and become cancerous.

WHAT IS CANCER?

I've made the analogy of cancers as rogue cells a few times already. Cancer is a generic term for diseases caused by overgrowth of cells that begin to invade or metastasize to other parts of the body. You can't "catch" a cancer cell, although there are certain viruses that you can catch that cause cancer. Cancer cells are derived from our own cells, usually because the fail-safes stop working. These are the fail-safes that stop cell division or cause cells to die via apoptosis when they start going rogue. These fail-safes break down because of mutations in DNA, genetic predispositions, or infection by certain viruses, such as the Human Papilloma Virus (HPV), which is linked to cervical and oropharyngeal (throat) cancer.[3] We are constantly bombarded with various mutagens that can lead to cancer. Mutagens are things like ultraviolet (UV) radiation or certain chemicals (carcinogens) that can damage or mutate our DNA and

---

3 Yes, HPV is a sexually-transmitted infection. Yes, you read correctly you can get HPV infections leading to cancer in your throat. That is a biology subject I won't be covering in this book. If you'd like to read more, the Kinsey Institute (kinseyinstitue.org) has great information on human sexuality.

cause cells to go rogue. Mutagens are even produced via our body's own chemical processes. The various fail-safes in our body work pretty well to protect our cells. However, people still die of cancer.

HOW DO PEOPLE GET CANCER?

I like to draw an analogy between getting cancer and being in an airplane crash. Both are relatively rare. Both have lots of fail-safes in place for protection. Both still happen. Cells going rogue and becoming cancerous have to escape many different biological controls. Screening for these rogue cells is what constitutes much of our yearly medical checkups and exams. If you are a woman, you probably remember getting a pap smear. A pap smear is when a brush is used to remove a few cells from the cervix, and the cells are looked at under a microscope to determine if they show signs of going rogue, evidenced by unstable DNA or genomes. If you've ever had a mole removed or a biopsy taken, the cells are examined under a microscope to see if they are in fact cancerous. Precancerous lesions often will have the margins cut to ensure that all rogue cells are removed. This is why yearly skin checks are so important. Melanoma (and many other forms of cancer) can be completely cured if it is removed from the body before it begins to spread.

Cancer spreading, or metastasis, is when rogue cells become aggressive and start to move throughout the body. If someone has breast cancer that spreads to the lungs and brain, they do not have breast, lung, and brain cancer. Instead, they have breast cancer that spread to other tissues. These cancer cells can retain parts of the original tissues, even in their new locations. For example, when I was in graduate school, one of the teaching samples we had for demonstrating parts of the brain was taken from an individual who died of melanoma. The tumor in the brain had pigmented tissue that looked like moles, not like normal brain tissue.

One of the reasons melanoma is such a deadly form of cancer is that in order to spread, it has to break through a layer of cells called the basement membrane. The basement membrane is a sheet of cells that forms a barrier and is one of our body's fail-safes. The name "basement" is a clue to where it is found: it is the bottom layer of skin. Generally speaking, most rogue cells can't break through this barrier. So instead, we get moles on our skin that are benign tumors. How do we know if a mole has become precancerous? If you think back to the literature you may have seen in your doctor's office or in the community, the warning signs are a mole that begins to grow or change. A mole may spread or become raised because it is growing, but it can't grow downward because it can't pass the basement membrane barrier. A mole that isn't growing, but then starts to grow is a warning sign. This is why if you notice any moles changing, it is a good idea to see your doctor to have them checked out. If the cells continue to mutate and become more aggressive, they eventually can break through this barrier and spread to other parts of the body, and the chances of the person surviving go down dramatically.

Maybe you've heard of genes that are related to cancer, such as BRCA1 and BRCA2. BRCA stands for **BR**east **CA**ncer, and mutations in these genes are associated with developing breast cancer. Everyone has BRCA genes. These genes produce a protein (we'll come back to the relationship between genes and protein in Chapter 6) that repairs damaged DNA. DNA damage is a pretty regular occurrence throughout your body and if not repaired can lead to cancer. Your BRCA genes *prevent* you from getting breast cancer by fixing DNA damage. BRCA genes are called tumor suppressor genes (we'll come back to tumor suppressor genes). If your BRCA gene is mutated for whatever reason, then your cells lose the ability to fix DNA damage. If this damage accumulates and other failsafe mechanisms are lost for whatever reason, cancer can result. This leads to an important point, one that we'll come back to in Chapter 7. Just because you have a mutated BRCA gene (for example, maybe you inherited a bad copy from one of your parents) *does not* mean you will get breast cancer. It means your odds for getting breast cancer are higher. There are multiple other cellular processes that protect you. For example, there are proteins that act like sensors for DNA damage. These proteins could still become activated and cause the cell to die through apoptosis before it can become cancerous.

One of these proteins is called p53, or the guardian of the genome. The protein p53 works to prevent cancer both by stopping the production of new cells and also by promoting cell death. The protein p53 can be thought of as a master regulator, or a conductor, for these various cancer-preventing processes in cells. Loss of p53 function is very common in cancerous tissues. Without p53, one of the key biological mechanisms of preventing cancer is lost, leading cells to go rogue.[4]

BRCA and p53 are both tumor suppressors. APC, a gene mutated in colon cancer, is also a tumor suppressor. If the function of a tumor suppressor is lost, it causes cells to go rogue by releasing

---

[4] For further reading see Rivlin, N., Brosh, R., Oren, M., & Rotter, V. (2011). Mutations in the p53 Tumor Suppressor Gene: Important Milestones at the Various Steps of Tumorigenesis. *Genes & Cancer*, 2(4), 466–474. doi:10.1177/1947601911408889.

them from fail-safes that prevent cancer. This is in contrast to oncogenes. Oncogenes are the opposite of tumor suppressor genes. Proto-oncogenes, or the normal unmutated form of oncogenes, help our cells to multiply. We need cells to divide and multiply to help replace damaged or worn-out cells. Not surprisingly, these cells are tightly controlled. Think of it like driving your car. A little bit of gas is desirable so the car will go, but you can hit the brakes to slow down. If a proto-oncogene becomes mutated and becomes an oncogene, cell division spins out of control. Going back to our car example, you've lost your brakes and the car is speeding faster and faster and impossible to manage.

WHAT MAKES CELLS GO ROGUE?

Damage to critical cells, either causing genes that prevent cancer to lose their protective abilities or genes that promote cell growth to lose their ability to be controlled, is what causes cells to go rogue. Several cancers, such as breast and colon cancer, have a genetic component. If you inherited a mutated form of a tumor suppressor gene that results in a protein that doesn't work very well, you are more vulnerable to getting cancer. One of the most critical aspects of developing cancer, and one that we are just beginning to understand, is the role of environmental factors in the development of cancer—after all, inheriting a bad copy of a tumor suppressor gene does not guarantee cancer, just as having all good and functional tumor suppressor genes does not mean you will not get cancer. Said otherwise, nature and nurture are both important when considering how people get cancer. The easiest one to understand is (UV) radiation and skin cancer. We know that exposure to UV radiation via sunlight causes DNA mutation and can eventually lead to skin cancer. This is why everyone is encouraged to wear sunscreen while outside.

What about chemicals in the air we breathe and the food we eat? Carcinogens are chemicals that are associated with causing cancer. For example, there is some evidence suggesting that air pollution, particularly pollution produced by gasoline-powered engines, is associated with development of breast cancer.[5] Cigarette smoking is another source of carcinogens—the National Cancer Society estimates that there are 69 different carcinogenic chemicals found in cigarette smoke. Some carcinogens you may be familiar with from your day-to-day life. For example, asbestos was used as insulation in buildings between 1930 and 1950 and in textured paints until the 1970s. You also may be familiar with how very expensive asbestos is to remove from buildings. Since inhalation of asbestos fibers can lead to serious medical conditions and cancers like mesothelioma, asbestos abatement, or removal of asbestos from a structure, is an expensive and time-consuming process to ensure it is completely removed without spreading it around. Another example is formaldehyde—it is used for preserving tissues and is also produced as a byproduct of our metabolism. Formaldehyde is produced naturally by our bodies and then through different processes in our bodies. Interestingly enough, going back to Chapter 3 and our discussion of natural and synthetic chemicals, both asbestos and formaldehyde are naturally occurring substances.

One reason finding a cure for cancer poses a challenge is because many different interacting processes have to go wrong. Said otherwise, we have to lose multiple fail-safes, and some of those

---

[5] There are many articles on this subject. Here is one of them: White, A. J., Bradshaw, P. T., & Hamra, G. B. (2018). Air pollution and breast cancer: a review. *Current epidemiology reports*, 5(2), 92-100.

fail-safes we don't completely understand. Many different kinds of genes interact with one another. Without getting into the details, in various signaling cascades associated with promoting or preventing cell signaling or apoptosis, there are multiple tumor suppressors and proto-oncogenes all working together. There also is your individual genetics, not only for tumor suppressors, but for all of your genes as well. How your individual genetics interfaces with your individual environment is not well understood.

Summary

As discussed at the beginning of this chapter, our knowledge of cell biology continues to grow exponentially as time passes. We find out new things all the time and revise what we knew previously. As we discussed in Chapter 2, this is the nature of science—new discoveries, new evidence, new ideas, and eventually new treatments. One of the most relatable applications of cell biology to our daily lives is cancer. Cancer is prevalent, and I know of no one who hasn't been impacted by cancer. Part of my motivation for writing a chapter on cancer was in memory of my in-laws who both died of cancer within six months of each other (see the picture at the beginning of this chapter). We take steps to protect our cells from going rogue with the choices we make, such as wearing sunscreen, avoiding carcinogenic chemicals, and seeing our doctor for yearly checkups. Our cells are the fundamental unit of life and orchestrate a wide variety of processes to keep us going. Where do cells get the energy for doing this? We will now turn to a discussion of energy in Chapter 5.

CHAPTER 5

Energy: From the Sun to Your Granola Bar

A FEW YEARS ago there was a viral video of recent Harvard graduates asked to explain where the mass of a tree comes from, a fundamental biological principle for explaining life on Earth. The viral nature of the video derived from the fact that none of the students were able to answer the question correctly. When the diplomats were posed this question, they contended that the tree's mass comes from nutrients in the soil. Dirt is "food" for plants, right? Sure ... there are nutrients that plants pick up from the soil, along with water. But that doesn't explain where trees come from. Plants get lampooned in biology classrooms on a regular basis for being lowly and boring, and teachers have a tendency to quickly gloss over the plant chapters or skip them altogether. However, without those lowly plants, there would be no usable energy available for much (but not all) of life on Earth ... and consequently, there would be no life.

Plants have the unique ability to undergo photosynthesis. Photo means light, and synthesis means creation. Plants create using light. The mass of a tree comes about because the tree can use energy from the sun to fix carbon, meaning turn carbon dioxide, a gas, into a chemical form. This chemical form then can be used to generate the structure of a tree and to produce glucose, which serves as a food source for both the tree and for other organisms. Think about that for a second. Plants use sunlight to transform gas into food, food that we rely on for our own survival.[6] This is also why when considering climate change, deforestation can be thought of as twice as bad for the environment. Not only are you taking away an organism that pulls carbon dioxide gas (which is a greenhouse gas) from the atmosphere, but when those trees are burned, the carbon dioxide that was stored by the tree is released into the atmosphere.

For a long time, it was thought that all life on Earth relied on the sun because of the ability of plants to use sunlight to make food for everyone and everything else. This is a nice example of something we discussed in Chapter 2: how science knowledge changes over time in light of new discoveries. When life was first found at the bottom of the oceans, far beyond the reaches of sunlight, we realized that not all life on Earth is dependent on the sun.

Whether the energy source is the sun, or derived from the Earth's core at the bottom of the ocean, the transfer of energy, and consequently the presence of life, is dependent on autotrophs. *Auto* means self, like an autobiography is a book about oneself, and *troph* is to feed. An autotroph self-feeds, or makes its own food. Plants are autotrophs—they use carbon dioxide in the atmosphere and sunlight to generate their own food. There is no need for plants to walk around or drive to the nearest supermarket; they just need a bit of sunlight and some carbon dioxide. It would be

---

[6] Plants also produce oxygen—this is different from the food I am mentioning here.

nice if we were autotrophs, too. Instead of cooking dinner tonight, I could just go lie out in my sunny backyard, breathe, and read a book. No more grocery bill! A plant's ability to undergo photosynthesis is a superpower—one that is foundational to energy and the first step in explaining how we go from the sun to fueling our own bodies.

Transfer of Energy.

Plants and other autotrophs are also called producers and have the greatest amount of energy. Producers are then consumed for their energy by another organism, say a bug. The bug gets eaten by a bird, and the bird gets eaten by a bobcat. Said otherwise, we move up the food chain. As we move up the food chain, energy is lost. Think of the energy available in the food chain as your monthly budget. What comes out of the budget first? Your monthly expenses. Rent/mortgage, utilities, food. What is left over is what you have available for going to the movies, shopping, or whatever else makes you happy. Producers also have a budget—they have to meet their energy needs first, plus have extra energy, since some energy is always lost as heat. What is left over is the energy available for other organisms to consume. So the bug has less energy available than that plant, because the total energy has decreased from what was originally produced by the plant. The bird and finally the bobcat have even less energy available to them. This is why a major solution to feeding humankind is for people to eat a plant-based diet. The total energy available by consuming plants is far greater and can feed more people than if everyone was consuming meat. A smaller food chain is akin to having fewer bills each month. There is less energy (money) going out, and therefore more to go around.

Another application of how things move between organisms that you may be familiar with is bioaccumulation or biomagnification. Have you heard the recommendation to not eat seafood like sharks, swordfish, some tuna species, and orange roughy? Particularly among young children and women who are pregnant or nursing? All of these species have one thing in common: they are top-of-the-food-chain predators and at the opposite end of the producers. So, sharks eat big fish, and those big fish eat little fish, and those little fish eat even smaller organisms like plankton … and those plankton in turn consume some kind of environmental pollutant like mercury.

Not only is energy being lost as it moves up the food chain, toxins like mercury can accumulate. Mercury is not broken down or excreted, so whatever an organism ingests, it remains in its system permanently. So the plankton might not have a very high amount, but the little fish will eat lots of plankton and slowly accumulate mercury until it has a modest amount. Then a big fish eats little fish that all have a modest amount and upon consumption now has a large amount. Then when something like a shark comes along and eats lots of the big fish, it suddenly has a massive amount of mercury. Then if we come along thinking to eat the shark, we are being exposed to potentially dangerous amounts of mercury. If you love seafood, eat lower on the food chain or limit how many predator fish you consume. Better yet, consider pro-sea life environmental protections that decrease pollutants like mercury in our waterways and seas.

The point I'm hoping to illustrate here is that we are all interconnected with other organisms on the planet, something we'll get into more detail later in this book in Chapter 9. We need to eat to obtain the energy needed to fuel our bodies, and how and what we eat has consequences both on our health and on the amount of energy available for everyone else. But how do we generate energy from our food?

Glucose and Fueling Our Bodies.

To recap, we absolutely rely on plants and other autotrophs because they can turn resources like sunlight and carbon dioxide into a form that we can use: glucose. Glucose is a carbohydrate, or as we discussed in Chapter 3, carbohydrates are also called sugars. The glucose in your bloodstream, or your blood sugar level, is how energy gets to your cells. Someone who has diabetes has a higher than normal blood sugar level. Normally, when your blood sugar rises (for example, after you eat breakfast), insulin is released to lower the blood sugar level by either helping cells absorb the glucose or stimulating the storage of excess glucose in either the liver or the skeletal muscle. Glucagon is another hormone that works in conjunction with insulin. Glucagon helps keep blood sugar levels from getting too low by stimulating the release of stored sugar in the liver. Type 1 diabetes occurs because the cells in the pancreas that make insulin do not work. This is different from Type 2 diabetes, which happens because cells don't respond to insulin (called insulin resistance) and blood sugar levels remain high.

What happens to excess energy? It gets stored as fat. The challenge with gaining too much weight is that it can lead to insulin resistance which then can lead to diabetes. Excessive blood sugar levels cause damage to the body and trigger additional weight gain, which, in turn, makes insulin work less efficiently causing high-sugar levels and more weight gain. It is quite a vicious cycle and difficult to break out of. This also is why dieticians often encourage you to eat anything with high amounts of sugar alongside a protein or a fat. Coupling something like crackers with peanut butter mitigates blood sugar spikes, makes it easier to process the sugar for energy rather than fat storage, and means you are less likely to have a sugar "crash" later that leaves you lightheaded. Something to be careful about when making choices in the supermarket is how various companies market high-sugar "healthy" foods. Sure, that package of fruit snacks may be low in calories and contain no fat, but it has an obscene amount of added sugar and will consequently spike your blood sugar or blood glucose levels. Added sugars are particularly nefarious since they are concentrated pure sugar in a form our bodies can regularly use. This is why long-distance runners will eat high-sugar "goos" to refuel mid-run. However, for the average person, the resultant spike in blood sugar is not desirable.

We often think about exercise when we want to lose weight. The interesting thing about exercise is that it helps you lose weight and avoid weight gain in two ways. First, exercise burns energy or calories. Any excess calories that you've consumed or that you are carrying around on your thighs gets used up. The second part may be less familiar to you. When you exercise, it helps your cells use glucose more efficiently. If you are using glucose more efficiently, it is less likely to get stored as fat helping to keep you lean.

Stress also can impact our weight because it causes insulin resistance and the release of additional glucose. If you think about this response in our hunter-gatherer ancestors, it makes sense. If you are in danger, you want more energy readily available so you can flee from the saber-toothed cat that is chasing you. The big deadline at work that is stressing you out has the same effect on your sugar availability, but without the physical requirements to burn the excess energy, it will ultimately lead to excess energy floating around and subsequent weight gain.

How Do Our Cells Use Glucose and What Does it Have to do With Beer?

We've covered how plants make glucose and how it gets into our cells. Once our cells have glucose, how does the glucose get converted into an energy form that is used by our cells? Said otherwise, how does our metabolism work? The energy form that our cellular machinery needs to "run" is called adenosine tri-phosphate (ATP). It starts with a chemical process called glycolysis. *Glyco* means sugar and *lysis* is breaking, so glycolysis is the process of breaking sugars.

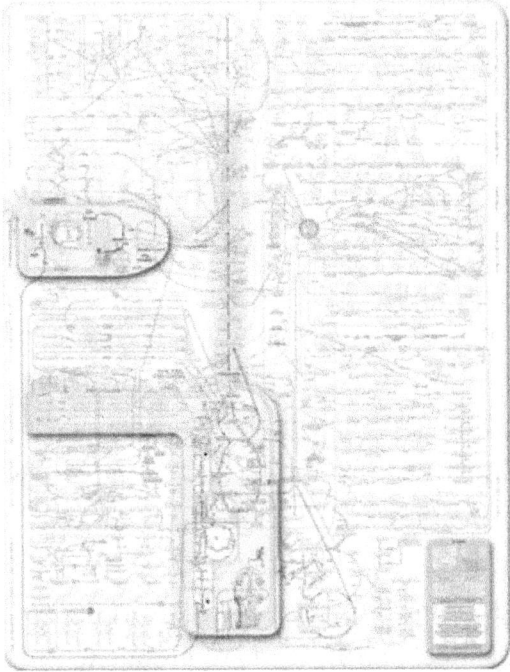

Figure 4. Diagram of all metabolic processes. Image used with permission from the International Union of Biochemistry and Molecular Biology.

Figure 4 is a classic image demonstrating how incredibly complex and interrelated the chemical reactions are that cells use to produce ATP. See the shaded part right in the middle of the diagram? That's glycolysis. It's shaded in because it is the central metabolic pathway. There are many other metabolic pathways that all feed into glycolysis. At best, you may have learned in your previous biology classes that glycolysis is the first pathway in cellular respiration or fermentation (we'll come back to those processes in a minute) and is used to convert glucose into pyruvate, an intermediate molecule that eventually will be processed further to make energy. At worst, you had to memorize all of the enzymes and steps between glucose and pyruvate. I had the misfortune of doing this twice in my educational career: in Advanced Placement biology in high school and in a college-level biochemistry class. How much do I remember off the top of my head? Not much. What I remember is what I just told you, that glycolysis is a central metabolic pathway—because that is the critically important part.

To make ATP from pyruvate, there are two different pathways. Which pathway is used depends on if oxygen is present or not. If oxygen is present, the pyruvate continues through cellular respiration, starting with a specific set of chemical reactions that make up the Kreb's Cycle and then continuing through the electron transport chain, which are the chemical reactions that actually produce ATP. Ever heard of the mitochondria? It is an organelle, or part of the cell, where the Kreb's Cycle and electron transport chain occur. This is why the mitochondria is often called the "powerhouse" of the cell. The mitochondria is where production of the fuel that cells need (ATP) occurs.

These chemical reactions explain quite a bit about our physiology. Let's start with why we inhale oxygen and exhale carbon dioxide. Glucose is made up of carbon atoms, and we don't need all of them. As part of the beginning process of the Kreb's Cycle, our body gets rid of the excess carbon atoms by converting them to carbon dioxide that we then exhale. Plants then take in that carbon dioxide and fix it to make more glucose that we'll consume later, going full circle. Neat! Why do we inhale oxygen? We need oxygen for the chemical reactions in the electron transport chain to work. Without oxygen, we can't produce ATP and will die. The reason cyanide and carbon monoxide are deadly to us is because they inhibit the role of oxygen in the electron transport chain. Arsenic is another poison that interferes with our generation of energy. Instead of impacting the electron transport chain, it blocks the Kreb's Cycle so these chemical reactions can't proceed. If the Kreb's Cycle is blocked, the electron transport chain is also blocked and therefore no ATP is produced.

The electron transport chain requires oxygen to make ATP. What if there is no oxygen available? How do organisms generate energy without oxygen? Although most of life on Earth lives in oxygen-rich settings, there are some exceptions. For example, a deep puncture wound or waterlogged soils can be anaerobic, meaning there is no oxygen present. In these cases, an organism will not undergo cellular respiration (the Kreb's Cycle and the electron transport chain) but will instead undergo the process of fermentation. Glycolysis remains the same since it doesn't require oxygen, but the resulting pyruvate will be processed to form ATP through fermentation.

There are two kinds of fermentation. If ATP plus ethanol and carbon dioxide are produced, this is alcoholic fermentation. If lactic acid is produced along with ATP, it is called lactic acid fermentation. Fermentation can occur in our bodies (such as during peak physical activity). We also can use fermentation to produce a variety of foods and beverages. However, if someone is poisoned with cyanide, why do they die if they can still undergo fermentation? Fermentation is not as efficient as cellular respiration. You don't get quite as much energy from the starting material. That is why fermentation is the only source of ATP in situations where there is a large supply of glucose present and why if you are making a fermented food or beverage, you need a large supply of sugar.

Now, where does beer come in? Alcoholic beverages are made through the process of alcoholic fermentation. The bubbles in beer or champagne result from the carbon dioxide that is generated during the process of fermentation. The alcohol is also a waste product from the production. To make fermented foods, we rely on an organism like yeast to break down the provided sugar to make bubbles and alcohol. At an *extremely* crude level, the bubbles are yeast farts and the alcohol is yeast pee. Single-celled organisms obviously don't have digestive or excretory systems, but these are the waste products that are produced. Why does bread rise? Carbon dioxide bubbles from the yeast.

One of my learning moments from my first experiences living away from home was accidentally making hooch in my college dorm. There was a bottle of old, previously opened apple juice in my refrigerator. I went to open it and all sorts of gas escaped. That was my first hint that I shouldn't drink the juice. In my naivety, I merely thought that it was strange. Then I took a sip. Yup. It was alcoholic apple juice and tasted nasty. Some kind of wild yeast or bacteria had entered the bottle when it was previously opened.

Since the sealed bottle didn't have oxygen, whatever organism was present started undergoing fermentation. Since apple juice is nice and sugary, the organism was able to survive by making energy for itself using fermentation. It certainly made an impression on me and is an example of fermentation in action. You can recreate this experiment for yourself or your kids by giving baker's yeast a solution with sugar and warm water and covering the top of the container with a balloon. As the yeast undergoes fermentation and uses up the sugar, the balloon will steadily inflate.

What about lactic acid fermentation? Lactic acid fermentation is used to produce foods with a sour taste. Yogurt, for example, is produced using a bacterium called *Lactobacillus* that ferments the milk sugars to produce lactic acid. Many other fermented foods get their characteristic sour taste due to lactic acid formation, including pickles, kombucha, kimchi, sauerkraut, and kefir. Fermented foods have an additional bonus of being less likely to spoil. Sourdough bread is unique in that it uses both types of fermentation. Alcoholic fermentation explains why it rises and can have an alcoholic tang, but the sour taste comes from lactic acid fermentation. A sourdough starter is a mix of live yeast and bacteria. A bit of starter provides the right kinds of yeast in a batch of bread to produce the characteristic flavor. San Francisco sourdough bread gets its characteristic flavor from a particular strain of wild bacteria, *L. sanfranciscensis*.

Summary

In this chapter, we talked about the superpower plants have that is the foundation for almost all life on Earth: their ability to use energy from the sun to convert carbon dioxide gas into a useable fuel. This explains why plants are a better source of energy than things that eat plants, and why if we want to feed the world, it is better to use plants than animals as our primary source of energy. We covered how energy gets into our bodies and our cells, and how too much energy or sugar in our bodies can lead to problems such as diabetes and obesity. We also examined our metabolism, focusing on the key chemical processes that turn glucose into an energy form that our cells can use called ATP. These processes explain why we inhale oxygen and exhale carbon dioxide, and why things like cyanide are deadly when ingested. In closing, we discussed how processes like fermentation can be used to produce ATP in low-oxygen environments, and how we as humans harness these chemical reactions to make food and beverages that are tasty and less prone to spoiling, like yogurt or beer.

This chapter is the last covering the chemical basics of biology, and now we are going to switch gears and spend the next two chapters on one of everyone's favorite topics: genetics. We first will cover the basics of DNA, sexual reproduction, and inheritance (Chapter 6) before turning to genetic privacy, technologies, and ethics (Chapter 7), some of the biggest issues we are facing as a society today.

CHAPTER 6

Mom Genes: DNA, Genetics, and Parenthood

I MENTIONED MY in-laws at the beginning of Chapter 4. Part of the tragedy surrounding their deaths was not only that they died in quick succession, but they both died while I was pregnant with my son. However, there was a certain sense of peace when several months after my father-in-law passed, my son started smiling for the first time and gave me exactly his grandfather's smile. My son shares DNA not only with me and my husband, but with his grandparents, great-grandparents, great-great-grandparents, aunts, uncles, and so on. People I love who have died live on not only in my memories, but in the genes they passed on.

This is part of the appeal of genetics, and for some in a potentially narcissistic way, the appeal of having biological children. Whenever I teach general biology classes and ask students at the beginning of the semester what they are most excited about, genetics is the top answer. One aspect that factored into my decision to biologically reproduce (said otherwise, have a baby rather than adopt or foster) was curiosity about the giant science experiment that goes into gestating a baby and meeting a creature that shares both my husband's and my genes. Genetics is an area of biology that is absolutely fascinating, and one that we'll be talking about in the next two chapters. This chapter covers the basics of DNA and genetics, while the next chapter will cover some of the biggest issues related to the genetic technology and privacy issues that we are currently facing as a society.

ROSALIND FRANKLIN AND DNA.

Have you ever heard of Rosalind Franklin? If you've read James Watson's memoirs on the discovery of the structure of DNA, you probably haven't heard very nice things about Franklin, or "Rosie," as he refers to her throughout the book, even though she didn't like that nickname. The story of how the structure of DNA was discovered is a classic example of how the process of science is influenced by culture.

As a female scientist in the 1950s, Franklin was not taken seriously by her male colleagues. This was during an age when women did not work outside the home, let alone pursue PhDs and do biochemistry research. Without Franklin, her male colleagues (James Watson, Francis Crick, and Maurice Wilkins) would not have been able to solve the structure of DNA when they did, let alone share a Nobel Prize. The missing piece of the puzzle to solve the structure of DNA required data that Franklin had generated, an X-ray crystallography image of DNA showing its double helical structure known as Photo 51. How did the critical image make it from Franklin to Watson and Crick? The history of what happened remains murky, but many suspect that the crucial image was given to Watson and Crick without Franklin's approval. In the subsequent manuscript describing Watson and Crick's proposition on the double helical structure of DNA, Franklin barely received an acknowledgement for her crucially important image, and, consequently, her work on the discovery of the structure of DNA has largely gone unacknowledged. Some speculate that she would have received the Nobel Prize alongside her male colleagues if she had been able to receive the award posthumously. She died of cancer at the age of 37 four years before the award was issued. It is thought that her work with X-rays contributed to her development of cancer.

Why was the discovery of the structure of DNA such a big deal? Because it allowed biologists to recognize DNA as genetic material. As we discussed in Chapter 3, DNA (and its kissing cousin, RNA) is a molecule that transmits information. Understanding how DNA is put together with various building blocks, those of A (adenine), T (thymine), G (guanine), and C (cytosine), is critical for understanding how it transmits information across time and space. Consider this analogy. This entire book is written based on 26 characters known as the alphabet. I can communicate an immense amount of information to you, because you and I both understand and can interpret these 26 characters joined together in different ways. This is the same principle that explains how DNA can be used to store biological information on everything from your hair color to your behavior. The letters are combined in a wide variety of manners to explain how they represent stored information. Molecular machinery in your cells is able to read that information and turn it into something else. For example, in Chapter 4 we talked about cellular machinery that fixes DNA damage or prevents cancer by preventing cell division.

Much like how this book is packaged up into pages and chapters to be easily digestible, DNA is packaged as well. DNA in a human cell contains 3 billion letters or base pairs (for reference, this chapter contains 22,687 letters and the entire book clocks in at 300,134 letters), and if you stretched the DNA end to end, it would be over six feet long. That is the DNA contained in ONE cell. You have approximately 37.2 trillion cells, and most of the cells (with the exception of gametes, the sperm and egg cells—we'll come back to those in a minute) contain your entire genome, which includes one copy of everything from your biological Mom, and one copy of everything from your biological Dad (plus a little bit of mixing—we'll come back to recombination as well).

The cells must organize DNA in a way that makes sense. Although some genes are only used once throughout life (for example, the gene that specifies development of a penis and testicles is only active during embryonic development, and without it, the default development is female), many others are in continual use. Proteins, the workhorses of each cell, need to be replaced, and DNA contains the instructions for how to build new proteins.

DNA is first organized into chromosomes. Humans have 23 individual chromosomes, which are paired to make 46 total chromosomes in each cell (except your gametes, which have 23 chromosomes). There are some exceptions, such as Down syndrome, which is when the individual has three copies of Chromosome 21. This also is why Down syndrome is called *Tri*somy 21. Within the individual chromosomes, DNA is wound onto special proteins that act like spools. DNA can be wound more or less tightly depending on which genes are located where. Genes are specific segments of DNA that code for particular proteins or RNA species that can act like proteins. Proteins are the workhorses of the cell and the macromolecules that do the "work" in the cell. For example, in Chapter 4 we discussed how p53, a protein, prevents cancer by stopping cell division and promoting cell death.

You can make an analogy between genes and recipes. If you go to the cookbook (genome) and pull out a recipe (gene), you then can write down that recipe (transcription, or making an RNA copy of a gene) and turn it into a tasty meal (translation, or turning the copy into a protein). Some genes are more accessible than others, and this is thought to be an important way the cell controls which genes are expressed and how often. The study of molecular processes that influence how easily genes are accessed and therefore expressed is a relatively new field known as epigenetics. We'll return to epigenetics at the end of this chapter.

REAL-LIFE X-MEN: DNA MUTATIONS, SEX, AND YOU.

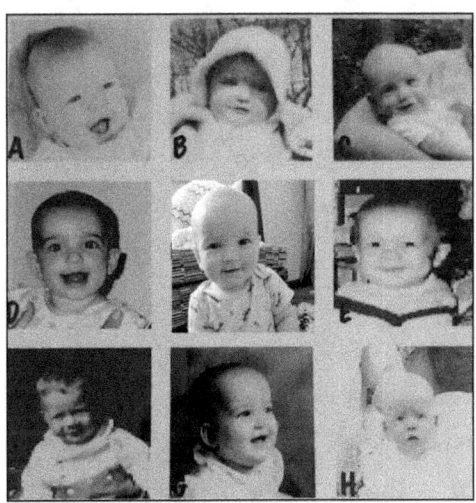

Figure 5. Name the baby game from my baby shower. Who do you think my son (middle panel) looks like?

The picture at the left (Figure 5) was for a game at my baby shower. The original game was to guess who is who based on the baby pictures of members of my extended family. I modified it to add in a baby picture of my son (middle). Who do you think he looks like? Can you pick out me or my husband? What about any of his grandparents and two of his great-grandparents? What traits do you see carried over generation to generation?

If we all share DNA, how come no one looks identical? Why are there traits that are unique to certain individuals? DNA changes over time through a variety of processes. This is important because genetic diversity is key to our survival (we'll come back to genetic diversity and conservation in Chapter 9). The more variety in our genetics, the more likely we as a species will survive. For example, if a new environmental challenge were to arise, if there is wide genetic diversity, it is more likely some members of the species can meet the challenge and survive. This is also how evolution works. We'll also return to evolution in Chapter 8.

How do we get genetic diversity? The first process is through random mutations in our DNA. A mutation is when the nucleotides (A,T,G,C) that make up DNA change. For example, perhaps an A is changed for a G. Since DNA is read by our molecular machinery in groups of three, and there can be many different groups of three that all lead to the same amino acid, which are the building blocks of proteins, sometimes these mutations have no effect and are called silent. Mutations that are significant enough to cause big changes in DNA can, in turn, alter how DNA is read by our molecular machinery and consequently which amino acids are used to build the protein. If the molecular machinery is reading incorrectly, then the resulting proteins are not built correctly either. Going back to our recipe analogy, it is as if your recipe became illegible, or was missing a key ingredient, and your resulting food was inedible. Cystic fibrosis, for example, is caused by three missing nucleotides leading to a nonfunctional protein and a devastating disease. But random mutations are not necessarily a bad thing. Although there is a potential for disease, they can provide advantages to an organism beyond increasing genetic diversity. For example, bacteria are able to overcome various anti-bacterial agents and survive due to random mutation. We will return to the genesis of super bacteria in Chapter 8.

Sexual reproduction is another way we generate genetic diversity. Genetic diversity is generated because of the contributions of genes from two different parents. There is a special subset of cells that sexually mature humans have called gametes. Gametes are special because they have half as much DNA as the rest of the cells. They have half as much DNA so that when a male gamete (sperm) meets a female gamete (egg), the resulting new entity (zygote) has the correct amount of DNA. If we use a money analogy and say that each of your somatic (body) cells are $1, then gametes are a 50-cent piece, or put another way, 50 cents + 50 cents = $1.

Gametes are produced through a special process called meiosis. This is different from the mitosis we talked about in Chapter 4 that is used for generating exact copies of cells. Since the goal is to get genetic diversity, it makes sense that with meiosis we would want to produce different cells. Not only does meiosis allow the body to produce cells with half as much DNA, but it allows for the generation of cells that are uniquely different both from each other and from the cells of the parents. This is why children can have traits not seen in either parent. The gamete produced is a unique mix of the DNA the parents originally received from their parents.

Let's back up a minute. In every cell in your body, there are two copies of the entirety of your DNA. One copy is from your biological mother and was present in the egg that was fertilized by your biological father's sperm (which is where your second copy came from). During meiosis, the DNA from your father and your mother can intermix to generate something completely new that then can be passed on to your children. During meiosis, it is completely random which gamete gets the new mixed up DNA or the unchanged DNA. Altogether, for *each parent alone* there are over

8 million possible combinations. This is how sexual reproduction can generate a huge amount of genetic diversity and why sexual reproduction is a major advantage over asexual reproduction, which produces a clone of an individual.

If it is extremely unlikely to get two identical individuals, where do twins come from? There are two kinds of twins: identical and fraternal. Fraternal twins are not identical. Fraternal twins come about when Mom releases two eggs during one fertility cycle (usually human females release only a single egg each month) and both eggs are fertilized by Dad's sperm and remain viable. Identical twins occur when one egg is fertilized, but the zygote splits in half. If the split is not complete, this can lead to conjoined twins, or twins that are physically attached to one another. There also can be rare events, such as the generation of a "pair and a spare," when two eggs are released and fertilized and one zygote splits. This leads to triplets made of a pair of identical twins and a third child that is fraternal. Identical twins are just that, identical, because they share all of the same starting genetic material. However, over the course of life, the environment can influence gene expression, and the resulting epigenetic changes can cause twins to become *less* identical over time. We'll come back to epigenetics at the end of this chapter.

Meiosis is a fantastic way to generate genetic diversity and, as already mentioned, confers some advantage to creatures that sexually reproduce. However, meiosis is not foolproof. The opportunity to exchange DNA and rearrange it leads to the likelihood of DNA ending up in the wrong place, or the gamete containing too much or too little DNA. Depending on the precise error, sometimes the gamete can still result in a viable offspring. For example, individuals with Down syndrome have an extra copy of Chromosome 21. Edwards syndrome, also called Trisomy 18, is caused by an extra Chromosome 18. Sadly, though, the majority of gametes with an incorrect amount of DNA will not go on to generate a viable embryo if fertilized. These meiotic defects are thought to explain at least 50 percent of all human miscarriages. It is estimated that approximately a quarter of all known human pregnancies end in miscarriage, although this number is likely an underestimate due to a lack of reporting or women who are unaware of the early stages of a pregnancy.

Mendel was Lucky.

Most of us have heard of the Austrian monk Gregor Mendel, the so-called father of genetics. Mendel was the first person to uncover some basic principles of how genes are passed on, or inherited, from parents to offspring. He happened to be lucky in the traits that he picked to study (pea plant height, pea color, and peas that are wrinkled or smooth), because these traits are a rare example of genes that are inherited in such a clean, simple fashion. Mendelian inheritance refers to genes that are inherited and expressed in a very particular pattern. There is a genotype (what genes you have) and a phenotype (the result of those genes, like your hair color). Genes can be dominant, meaning you can almost always see their phenotype, or recessive, meaning you will only see the phenotype if there is not a dominant copy present (remember, you have two copies of all of your genes, one from each biological parent). Punnett squares are a tool that you can use to track how genes are inherited and make predictions (I've got a real-world example of Punnett squares and inheritance coming up).

There can be some exceptions to the dominant/recessive rules. For example, there is incomplete dominance where an intermediate phenotype will be displayed if the individual has a dominant gene from one parent and a recessive gene from the other. The classic textbook example is the generation of a pink flower by an individual plant with a gene for red and a gene for white. Pink is generated by the mix of red and white. Codominance means regardless of what gene versions you have, both will be expressed. Again, turning to the classic textbook example, codominance is often explained in terms of blood types. The "A" and "B" we hear about when talking about blood types refers to the types of sugars that are on blood cells. A person with Type A blood has A sugars on their blood cells. If a person has a gene for A blood and a gene for B blood, they express both genes and have A and B sugars on their blood cells for Type AB blood. Type O blood means there are no sugars on the blood cells. Whether blood type is positive or negative refers to whether or not the Rh factor, a type of protein found on blood cells, is present.

When my son was born, we had a rather frightening but biologically interesting first few days. My son and I are ABO incompatible, and how and why that happened is an interesting exercise in inheritance and blood typing. We knew from my prenatal blood tests that I have Type O blood. This means I have no sugars on my blood cells and my genotype is O/O. If I were to need a blood transfusion, it would have to be Type O blood; otherwise my immune system would recognize the blood as foreign and "attack" it. This is why individuals with O negative blood are universal donors. There are no sugars or proteins on the blood that would react with someone else's body. My son, however, does not have Type

O blood; he is Type A. My immune system reacted to his blood because the A sugars were recognized as a foreign invader and were causing the destruction of his blood cells. The breakdown of blood cells was leading to a buildup of a protein called bilirubin, which is yellowish in color and can lead to jaundice, or yellowing of the skin and eyes. To help break down the excess bilirubin, my son spent the second and third days of his life on a BiliBed (Figure 6), a phototherapy system that uses UV light for the treatment of jaundice.

Figure 6. My one-day old son on his BiliBed to treat jaundice.

How did I have a son with Type A blood if I am Type O? Well, we have to blame my husband. If I'm Type O and if blood types are codominant, that means I must have the genotype O/O and could only give my son an O gene. Since he is Type A blood, and we know that my husband *must* have passed on an A gene, making my son A/O. So, if we were to draw this up with a Punnett square, it would look like what I've drawn in Figure 7.

My genotype is shown on the top, and my husband's is on the side. We know he has an A gene that he passed on to my son, but we do not know what his second gene type (or allele) is or his blood type. If we wanted to predict if I will be ABO incompatible with any future children, we can use the Punnett square to show all possible gene combinations (represented by the squares). I can only pass on an O, which is why there is an O in each square (the letter on top drops to all squares below). However, without my husband's genotype, we only know the chance of it happening again is at least 50 percent (chance and probability as they relate to genetics is something we'll talk about in the next chapter). Without knowing my husband's blood type, we can speculate that he is either Type A or Type AB since he has an A gene for sure, and since it is codominant, it will be expressed. In terms of his genotype, he could be A/A (two copies of A, meaning we would predict that 100 percent of our offspring will be ABO incompatible with me), A/O (meaning we would predict that 50 percent of our offspring will be ABO incompatible with me), or A/B (meaning again, that we would predict 100 percent of our offspring will be ABO incompatible with me).

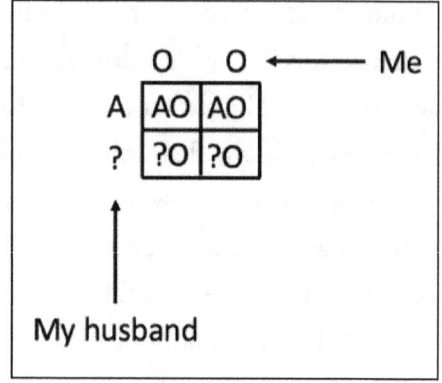

Figure 7. A Punnett square depicting possible blood types of our offspring.

Pedigrees can be useful tools for tracking genetic patterns across generations and helping to determine how a trait or a disease oftentimes is transmitted between generations. For example, a pedigree can be useful for determining if a disease or condition is sex linked or not. Sex linkage is another interesting inheritance phenomenon. Remember, humans have 23 pairs of chromosomes, and one of those pairs is the sex chromosomes. Generally, as there are exceptions, males have an X and a Y chromosome and females have two X chromosomes. The Y chromosome has a single function: to specify male during embryonic development. Although there are some other genes on the Y chromosome, most are involved in sperm production and not relevant to overall health. The X chromosome, however, carries more than 1,000 protein-coding genes, including genes that are responsible for everything from color blindness to diseases like hemophilia. Since males only have a single X chromosome, they are more likely to be affected by X-linked traits. They only have one chance, so to speak, to get a normal, or wild type, version of the genes. Going back to pedigrees, if a disease or condition were X-linked, we would see more males affected by it than females.

For a non-sex linked disease, the pedigree will vary depending on if it is a recessive or dominant gene that leads to the disease phenotype. Let's consider albinism. Albinism, or the lack of pigment, is recessive. In order for someone to have albinism, they must have two recessive genes. On a pedigree, you would only see affected individuals from parents who either are affected themselves or are carriers. A carrier is an individual who has one wild type and one mutant copy of a gene. They themselves are unaffected but can still pass the gene to their offspring. This applies to sex-linked traits as well. A woman can carry a mutant form of the gene controlling color blindness and not be color blind herself.

She can, however, have a color-blind offspring. With an autosomal (autosomal referring to a gene on an autosome, or non-sex chromosome) dominant trait, since it will be expressed if present, you would see many more individuals affected on a pedigree with no bias toward either sex.

What About the Epigenome?

I've been privy to discussions about whether or not to continue to teach Mendel in biology classrooms, or at least to minimize discussions of Mendel and Mendelian Genetics. Why? This stems back to the introduction to the last section: Mendel was lucky in the traits that he picked. The majority of genes are not inherited in such a straightforward fashion, and the patterns we see from pedigrees can be very difficult to interpret. In fact, as we'll get into in Chapter 14, strict adherence to the one gene/one phenotype idea, particularly when discussing diseases that are common among certain ethnic and racial groups, can actually fuel white supremacist thought. Genes and the proteins they code for rarely operate in isolation but are a part of a vast network. For example, a single gene can have many different phenotypes because the protein it encodes is important in multiple biochemical pathways. Remember the picture from Chapter 5 of the chemical reactions that are used by our cells to produce ATP? The picture shows how our biochemical pathways are complicated and interrelated with one another. There also can be oddball mutations that cause the phenotype to be different from the expected genotypes. For example, going back to our blood cell example, there is an extremely rare genotype that causes A or B sugars to not attach to the blood cell, giving the individual Type O blood ... even if genetically, they have the genes for A or B.

Epigenetics is a relatively new field beginning in the 1990s. The "official" definition was not even generated until 2008. Epigenetics refers to heritable changes on DNA that are *not* due to changes in the actual DNA sequence. These heritable changes in DNA result in changes in gene expression due to modifications on DNA leading to changes in accessibility. The fascinating part about our epigenome is that it is subject to change. Although the DNA we receive from our parents is unchanging, how that DNA is expressed can vary over the course of our lives. When we consider twins, this explains how they can go on to have very different health histories over the course of their lifetimes even if they have the same genes. Their epigenomes change over time, causing them to become less identical as they age. Our easily changed epigenome can be good or bad, as we can undo damaging epigenetic changes and this could be a viable treatment for diseases such as cancer; however, the easily changeable nature of our epigenome also means our genomes are more susceptible to bad epigenetic changes as well.

These epigenetic changes can even be passed on to our offspring. For example, intergenerational trauma or stress can be passed on from parents to offspring. One landmark study followed up on children who were *in utero* during the Dutch Hunger Winter in the 1940s. During World War II, the Dutch attempted to help Allied forces by interfering with railroad operations used by the Nazis to move troops. The Nazis then punished the Dutch by blocking food supplies and causing a famine that lasted until the Netherlands were liberated the following spring. These children went on to have altered metabolism and stress responsiveness as a result. They were more likely to weigh more, have high blood triglycerides, and have elevated cholesterol, and they were more likely to die younger. Evolutionarily this makes sense. If Mom faces significant stress, this prenatal programming can help prepare the offspring. If Mom is dealing with famine, it makes evolutionary sense for her offspring to be able to pack on the pounds for their survival.

One of the most interesting studies I've ever read about epigenetic changes and inheritance was done by a group at Emory University in Atlanta, Georgia, headed by Dr. Kerry Ressler.[7] In the study, researchers fear conditioned male rodents. Fear conditioning works the same as classical conditioning, except the organism in question begins to associate a stimulus with something fearful, such as pain. In this study, male animals associated a mild foot shock with the smell of almonds. Eventually, anytime the male animals would smell almonds, they would freeze in fear. These fear-conditioned male animals were bred with non-fear-conditioned females. The offspring were raised to adulthood and exposed to the almond smell. Guess what? The offspring, in spite of never coming into contact with their fear-conditioned father or being fear conditioned themselves, *had the same fear response* to the almond smell. The behavior was passed down through epigenetic changes in the sperm.

SUMMARY

In this chapter, we covered a wide range of topics related to the basics of genetics and inheritance. We discussed the discovery of DNA and how DNA can store and transmit information across generations. We talked about mutations and meiosis-generated genetic diversity and explained why children do not look exactly

---

[7] Dias, B. G., & Ressler, K. J. (2014). Parental olfactory experience influences behavior and neural structure in subsequent generations. *Nature neuroscience, 17*(1), 89.

like their parents. We learned how Mendel was lucky in that he studied traits that are passed down in predictable patterns across generations, and that modern genetics suggests that these are the exception and not the rule when it comes to considering human genetics. We closed with a brief introduction to epigenetics and how changes in the expression of DNA can lead to significant changes over the course of our lives. We discussed a few examples of genetics at play in our daily lives. But what about genetic technologies? What about that at-home genetic testing kit you saw at the store, or that news broadcast you saw on three-parent babies, or all those pesky GMO (genetically modified organism) labels you see at the store? We'll now turn to a discussion of genetic technologies and the ethical quandaries and questions surrounding them.

CHAPTER 7

Oh, Brave New World

"We want to give your child the best possible start. Believe me, we have enough imperfection built in already. Your child doesn't need any more additional burdens. Keep in mind, this child is still you. Simply, the best, of you. You could conceive naturally a thousand times and never get such a result."
-**The geneticist in the movie** *GATTACA* **to parents seeking genetic technology to conceive their second child**

IN THE OPENING quote and chapter title, I mention two of many pop culture references to technology and society. *Brave New World* by Aldous Huxley is a dystopian novel where the government has used technology to manipulate and control society. In the book, people are genetically engineered to be part of certain classes based on their intelligence. *GATTACA* is a 1997 dystopian movie where genetics are used to determine a person's place in society. *GATTACA* has similar appeal as the original *Star Trek* episodes in that close attention is paid to current technology and science and future projections ... and several aspects have since converted from science fiction to science fact.

This is, without a doubt, an extremely exciting and interesting time to be alive when it comes to genetic technology. Many of you reading this probably remember when the first human genome was sequenced—when the order of all nucleotides was uncovered for the first time— between 1990 and 2003 at the cost of $3 billion Whole genome sequencing is now available to the average consumer at a cost of a few hundred dollars. Pause for a second. In the last 20 years, the price of whole genome sequencing has dropped from the *billions* and *taking more than a decade to complete* to costing in the *hundreds* and *taking a few weeks*. In the early 2000s, the genetic technology community was abuzz with discussion of the $1,000 genome and how it would revolutionize the face of modern medicine. We could easily sequence and personalize treatment for patients! Now, sequencing is cheaper, faster, and more readily available than ever before. You can go online and get ancestry sequencing done for $99 with 23andMe.com or ancestry.com. You can also get health sequencing for $149-$199.

Genetic Sequencing: Empowering Good Health Decisions or Scaring the Pants off Everyone?

Before you run out to your nearest grocery store and buy an at-home genetic test kit and get your genome sequenced, let's pause to consider the benefits and potential consequences of at-home genetic testing. For one, if you are a biology nerd like me, knowing your genotype is really interesting. You read all about my exploits in figuring out genotype and blood typing in Chapter 6. Second, if you are interested in family history, knowing your genetic ancestry can be fascinating. And third, there also are numerous health benefits. Celiac disease is a fairly common immunological disease caused by sensitivity to gluten, a protein found in wheat, rye, and barley and cross-contamination in oats. Celiac disease also is notoriously difficult to diagnose because it has a high false negative rate, meaning if you are not consuming adequate amounts of gluten during the test, your test will be negative, even if you do actually have celiac disease. This has led to many individuals suffering with the disease but going undiagnosed for years. Now, you can get a genetic test for the genes that are most likely to cause celiac disease.

Other genetic testing can be instrumental in choosing which medications to prescribe. For example, some individuals have a genetic mutation in a protein called SERT that is involved in serotonin transport. With this mutation, patients will not respond to front-line antidepressants or anti-anxiety medications such as sertraline or fluoxetine. If you know anyone who has needed to be on an antidepressant or anti-anxiety medication, finding the right medication can be extremely challenging, and this information can save months of heartache for patients as they wait to find the right lifestyle and pharmacologic management strategies for their symptoms.

Knowing your genotype can be useful for making long-term health decisions. We know of certain diseases that are associated with particular genes. In Chapter 4, we talked about genes that when mutated are associated with an increased risk of cancer. With this information, perhaps if you know you carry mutated forms of genes associated with the development of colon cancer, you may wish to start getting colonoscopies earlier in life. Or, perhaps you've had a family member with colon cancer and you get tested and find out you don't carry the gene variants that are associated with the development of colon cancer. Then, maybe you do not need to start having colonoscopies before age 45. Information is power, right? The more you know, the better informed decisions you can make about your health.

However, the problem with paying too close of attention to our genetics is that it can lead to a common fallacy called deterministic genetics. If I have XYZ gene mutant, then I will absolutely have XYZ phenotype, too. This is the line between knowing your genetic information and using that to make empowered medical and life decisions, and letting it ruin your life. This was a common theme in *GATTACA* as well. One of the main characters had a high probability of heart problems and would not run due to concerns about having a heart attack. As we discussed in Chapter 6, your genes and proteins are part of very complicated pathways, and this is far more common than diseases that are caused by single genes. We hear more about diseases that are caused by single genes because they are much easier to understand. There is also the influence of epigenetics and the environment. If you have a certain gene associated with a certain disease, it only means you are *more likely* to have that disease. It does *not* mean you will get it. As discussed in Chapter 4, part of the reason cancer is such a hard disease to treat is because the process of getting cancer is complicated, and we don't fully understand all the routes into its development. I've heard the analogy of a loaded gun before, where your genes "load the gun," but epigenetics (environment) "pulls the trigger". Understanding how genes load a gun and how epigenetics can result in "pulling the trigger" is not well understood.

Here is a hypothetical example about the benefits and drawbacks of genetic testing. Say you have two kids and you sequence the genes of both kids using an at-home testing kit that includes health-related genes, including the BRCA genes that are implicated in developing breast cancer. You find out that your 3-year-old daughter has the BRCA mutation. What do you do? Worry about it for the next 30 to 40 years waiting to see if she'll develop breast cancer? What impact will it have on the child knowing that she is more likely, but not definitely going, to develop breast cancer? You can see how this could have a significant negative impact on the quality of life for the entire family. Again, this could lead to empowering decisions (like not waiting to have kids or getting regular breast exams), or it could lead to taking drastic measures to mitigate the risk of cancer, such as getting a preventative bilateral mastectomy (removal of the breasts).

The other ethical issue here is that we share our genes with our families. By getting yourself sequenced and choosing to release those results to another person, you are not releasing only your own health information. This is information about your parents, siblings, and children as well. Maybe you are worried about getting a life-threatening disease that runs in your family, so you get sequenced and find out you have a disease-causing gene variant. But maybe the rest of your family doesn't want to know. In 2013, a European group of scientists published the HeLa genome. HeLa is a cell line, or specific type of cells, derived from the cervical tumor obtained without consent from a black woman named Henrietta Lacks in the 1950s. HeLa is famous not only because of the ethical issues associated with its use and generation, but because it was the first time cells could be grown outside the body. It was used to generate a wide variety of scientific breakthroughs, including the development of the first polio vaccination. When the genome was published, the family protested, stating that since they were related to Lacks, this was potentially sensitive genetic information about themselves they did not want available publicly. The genome was later taken down and now is only available to qualified researchers. Another point to consider, although it is illegal in the United States to discriminate based on genetic information, is that the information could be misused in some manner.

Speaking of governments and genetic sequencing, it is common in the United States for mandatory genetic testing to be performed on newborns. At first glance, this makes sense. For example, in the state of Colorado (where I live), there are 37 different disorders tested for at birth, all of which are actionable with early treatment that is available and helpful. In the case of diseases such as PKU (phenylketonuria), the earlier the disease is diagnosed and addressed, the better the health outcomes are for the patient. So, did I consent to have my son sequenced? Of course! However, it didn't fully sit comfortably with me either. Does the state store that information? Is anything else tested for? Is there other important health information that could or was generated that I need to know about? All I was told by our pediatrician is that his testing was "good, nothing to worry about." This is another interesting ethical quandary. Do I have a *right* (and eventually, does my son have that *right*) to know this health information? Is there something I need to know about to take preventative action? Or should it be withheld, because too many people lack an understanding of genetics and probability, and these results would cause undue fear? Where is the line?

The advent of fast genetic testing has raised other issues as well. For example, according to United States patent law, it is possible to patent genes—assuming the investigator in question is the first person to identify and isolate the gene and describe the protein coded for by the gene. We all *have* these genes. How can you patent something that we all have? This gets even trickier when biotech companies generate tests to see if you have certain disease-causing variants, then charge exorbitant fees for the use of the test results. Since they own the patent for that gene or gene variant, they have a monopoly on generating the test as well. Obviously, they need to recoup monies spent on research and development (which is very expensive, a topic we will return to in Chapter 13), but there still lies the underlying question about patentability of something that each of us uniquely possesses.

Frankenfoods or Just Speeding Things Along?

Selective breeding has been around for centuries. Selective breeding is the process of breeding two organisms with a desirable trait to generate more offspring with said desirable trait. It is the process that eventually led to domesticated dogs and cats, as well as specific breeds. Small dogs were selectively bred together to generate more small dogs until eventually we ended up with Chihuahuas. This process also works with plants. For example, in the late 1800s, hybrid cereal grains were generated by intentionally breeding rye and wheat.

In the 1980s, genetic engineering was introduced to this process. Instead of breeding plants and hoping to get the desired traits, scientists now could insert desired genes into the plants. It also was possible to generate the first transgenic plants. Transgenic refers to an organism that has DNA from another species. This is a bit more common than you might realize. For example, there are several viruses that, upon infection of cells, inject their DNA into your cells, and that viral DNA is then incorporated into your genome. Generally, this isn't a big deal, but this is how some viral infections (like Human Papilloma Virus, or HPV) can cause cells to change and eventually lead to cancer. Have you ever seen GloFish before? These are genetically engineered (GE) fish with a gene from a jellyfish that makes them glow in the dark. Genetically modified (GM) bacteria are given the human gene for insulin and go on to produce insulin that can be used for people who are diabetic.

A common objection to GMs, however, lies with genetic modifications in the plants we eat. The first genetically modified plant associated with human consumption approved in the United States was a tomato variety with a longer shelf life. That was in 1994. Now, the majority of all soy, corn, and cotton in the United States are from GM plants. Part of this wide adoption is because using GM crops decreases pesticide usage, increases crop yields, and overall makes for a better bottom line for farmers. If it sounds too good to be true, it usually is. GM crops come at a cost. For example, Roundup Ready crops are the most common GM crop in the United States. Roundup is a common herbicide, or chemical, that kills plants. You may have some in your garage and use it to clear the weeds out of cracks in your sidewalk. The problem with Roundup is that it is linked to the global collapse of amphibians.[8]

Are GMs themselves bad? Likely not. The science is still new enough that it is good we are paying attention to and studying GM crops, while also being thoughtful about their usage. However, in the case of Roundup Ready crops, it isn't the fact they have a GM

---

8 Relyea, R. A. (2005). The lethal impact of Roundup on aquatic and terrestrial amphibians. *Ecological applications*, *15*(4), 1118-1124.

that is concerning, *but the nature of the modification.* Since these crops can resist Roundup, more people are using Roundup and our food is being sprayed with it. We know Roundup is bad news for the environment.

I've also found that consumer decisions become quite difficult when making selections at the grocery store. Labeling something as "non-GMO" is synonymous with "this will cost you more money." Non-GMO milk? That doesn't make any sense. Milk isn't alive. Although it could have DNA in it, it certainly isn't GM DNA. Genetically modified cows are not super common yet, although a study in 2017[9] used a new gene editing tool called CRISPR to engineer a few cows to be resistant to bovine tuberculosis. Should you pay 50 cents more for the non-GMO milk? Probably not. It's a marketing scheme to get you to buy a more expensive product.

Since many GM crops lead to improved crop yields, enhance drought or flood resistance, or add in vital micronutrients (for example, Golden Rice that contains extra Vitamin A), these crops help feed more people using fewer resources. They also can help make healthier foods more readily available for the general public. Some hypothesize it might be possible to engineer peanut plants without the protein some are deathly allergic to. This is, of course, balanced with Roundup Ready crops that have detestable qualities. I hope in the future we see more education about *which* GM was used in a particular product and have this linked to safety research. This will help inform consumer choice instead of breeding hysteria over "frankenfoods," or driving predatory marketing

---

9 Gao, Y., Wu, H., Wang, Y., Liu, X., Chen, L., Li, Q., ... & Zhang, Y. (2017). Single Cas9 nickase induced generation of NRAMP1 knockin cattle with reduced off-target effects. *Genome biology, 18*(1), 13.

schemes that scare consumers into buying anything labeled as non-GMO.

Given the prevalence of dystopian stories in our society, and as discussed in Chapter 2, the image of scientists portrayed as the bad guys or profiteering without a care for society, it is not surprising that fears about GM crops can take hold. In fact, studies show that the general public is far more likely than scientists to have reservations about the consumption of GM foods.[10] There is a need for more dialogue between people and scientists so that fears can be identified and research and information can be passed along to consumers in a manner that makes sense.

In closing for this section, I'd like to make one thing very clear: I am neither pro- nor anti-GM organisms. What I am is pro-education, and my hope is that this section was helpful in fleshing out the pros, cons, and gray areas around GM usage and that you feel more empowered in your consumer choices. Now, we will turn to an examination of some exciting, yet ethically dubious technologies: preimplantation genetic diagnosis (PGD), de-extinction, and three-parent babies.

A Quick Foray into Really Exciting, Yet Ethically Dubious, Technologies.

Preimplantation Genetic Diagnosis (PGD).

The opening quote of this chapter, from the movie *GATTACA*, was from a scene depicting PGD. The geneticist goes on to say that the four candidate embryos shown to the prospective parents had been screened already for "potentially prejudicial conditions,"

---

10 Funk C, Raine L (29 January 2015) "Public and Scientists' Views of Science and Society." Pew Research Center.

which included innocuous traits such as myopia (nearsightedness) and baldness. I have worn glasses since I was a young child ... so in this fictional dystopian world, I would have been selected out. We spent most of the early part of this chapter diving into sequencing and genetic probabilities. PGD allows for human embryos to be sequenced and intentionally selected for certain traits.

How does PGD work? Well, it starts with *in vitro* fertilization. *In vitro* is Latin for "in glass," so *in vitro* fertilization is fertilization that occurs in a dish (likely plastic, not glass) rather than in the body (*in vivo*). Once fertilization occurs, the embryo goes through mitotic divisions generating many identical cells. Geneticists can pluck off one of these cells and sequence the DNA inside. There are certainly medical benefits for doing this. Perhaps the parents are concerned about passing a debilitating, life-threatening illness on to their offspring, like cystic fibrosis. Embryos could be screened for the gene that causes cystic fibrosis, and those that have it can be screened out. Only the embryo(s) not expected to not have cystic fibrosis based on its genetic sequencing would be implanted into the mother.

Think of the heartache and pain this could save families. But going back to our discussion earlier in this chapter and in Chapter 6, we saw how our genes are only one piece of the puzzle. It is fairly uncommon for particular diseases to be linked singly to genes. Most genes interact in complicated pathways and/or require environmental triggers to cause progression to a disease state. Even if we know certain genes are linked to certain diseases, oftentimes it only confers a change in risk, not a definitive diagnosis, that a person will develop X or Y disease. With imperfect information and the thought of discarding unwanted fertilized embryos having ethical and moral considerations of its own, do these considerations outweigh the possible benefits to PGD? Are there examples where PGD is appropriate, for example in the case of diseases like cystic fibrosis that do have a clear genetic cause, which is how it is currently used?

The other critical ethical consideration is that this could pave the way for designer babies. Already with our existing technology, it would be fairly simple to choose a male over a female child. Things like height and eye color are a bit more complex but could be in the realm of possibilities in the future. Genes associated with improved athletic performance have been identified.[11] Although like many other traits, athletic performance is likely a mix of genetic predispositions with environmental factors. *GATTACA* was premised on designer babies creating an elite upper class, whereas people conceived the "old-fashioned way" were relegated to an underclass. As our ability to sequence our genes quickly and effectively continues to grow, as well as our knowledge about genes and their phenotypes, it is worth a close consideration of how to avoid some of the more dubious aspects of this movie becoming a reality.

The Science of De-extinction.

Extinction is when all members of a given species have died out. It is possible to be completely extinct, or extinct in the wild. Extinct in the wild means that a species is no longer found in the wild, only in human care such as in zoos or aquariums. Extinction, and the opposite process, speciation (more on that in Chapter 8), are

---

[11] Guth, L. M., & Roth, S. M. (2013). Genetic influence on athletic performance. *Current opinion in pediatrics*, 25(6), 653–658. doi:10.1097/MOP.0b013e3283659087.

normal biological process. The problem is human-driven habitat destruction (habitat being natural areas where animals live) has caused the background or normal extinction rate to rise 100 times and the entry into what many consider to be a mass extinction event.[12] Although there are conservation efforts in National Wildlife Refuges and other areas, such as Species Survival Plans organized by accredited zoos and aquariums for facilitating the breeding of endangered animals and setting aside land for them, what about the animals we've already lost?

Enter the science of de-extinction. De-extinction is using genetic technologies, such as CRISPR, the DNA editing tool mentioned earlier in this chapter, to edit the genes of an existing animal to be like those of a related animal that has gone extinct. If this makes you think *Jurassic Park*, a 1993 film where scientists used preserved DNA to bring dinosaurs back to life, you are on the right track. Rather than a mosquito trapped inside amber, scientists are using DNA found in species like wooly mammoths frozen in permafrost. The DNA is incomplete but gives scientists enough of a template to modify elephant DNA to look more like a mammoth. The idea would then be for an elephant to gestate a baby that has the modified genes.[13]

Now, if you've seen *Jurassic Park*, you know that de-extincting dinosaurs didn't go so well for the characters in the movie. The entrepreneurial idea of Jurassic Park seems sound. People go to zoos to see animals from the other side of the world; imagine what they

---

12 Ceballos, G., Ehrlich, P. R., Barnosky, A. D., García, A., Pringle, R. M., & Palmer, T. M. (2015). Accelerated modern human–induced species losses: Entering the sixth mass extinction. *Science advances*, *1*(5), e1400253.

13 For a complete story, I highly recommend Beth Shapiro's *How to Clone a Mammoth: The Science of De-extinction*.

would pay to see a real-world mammoth! What could we learn about their behaviors and physiology? Additionally, since humans are responsible for such massive rates of extinction, do we have a responsibility to undo some of the damage we have caused? Do we have an obligation to bring back animals like the Rabbs' fringe-limbed tree frog, which is thought to have recently gone extinct? The flip side of the argument, however, looks at whether or not it is okay to bring animals back if they have nowhere to go besides captivity. If their habitat is destroyed, or in the case of animals that have been extinct a long time and habitat is potentially dramatically changed, is it ethical to bring them back? If we bring a species back and try to reintroduce it without making major changes to restore habitats and prevent future loss, is that animal going to end up going extinct again? This is particularly relevant when considering animals that went extinct due to disease, such as the wide variety of frog species that went extinct due to an infectious fungus called chytrid. Many other frog species remain threatened by chytrid. De-extinction is a relatively new field, and who knows what we will see in the next few years.

Three-Parent Babies.
Most of us have two biological parents. The DNA in the nucleus of our cells, or our nuclear genome, comes from both of our biological parents. However, we also have a second set of DNA separate from the nuclear DNA that is found in the mitochondria. We talked about mitochondria in Chapter 5 as the powerhouse of the cell where our ATP energy is produced. The mitochondria also are unique in that they have a separate genome we get solely from our mother. There are several genes that lead to devastating diseases found only in the mitochondrial genome, such as Leigh's

syndrome, which causes neurodegeneration and death in early childhood. Since the genetic defect is found in the mitochondrial DNA, and mitochondrial DNA comes from the mother, it is very likely that she will pass the mutant gene onto all of her offspring. The mother may not be affected because there are multiple mitochondria in each cell, only some of which will have the devastating gene mutation.

How are three people involved? The biological father continues to contribute only sperm (parent 1). The first biological mother with the mitochondrial mutation contributes *only* her nuclear DNA. Her nucleus is removed from her egg, leaving behind the mitochondria and genetic material that contains the mutation (parent 2). A second woman donates an egg with mitochondria that is healthy (parent 3). Her egg's nucleus is removed and parent 2's nucleus is placed in the egg from parent 3, and then the final engineered egg is fertilized with the sperm from parent 1. The fertilized egg is then transferred into the mother. The first child from this procedure was born in 2016.

Three-parent babies are a potentially exciting avenue for helping parents who are unlikely to have healthy children otherwise. Due to the genetic manipulation of human gametes to generate a GM embryo, this process is, at least of the writing of this book, still illegal in the United States. It has become legal in other countries, such as in the United Kingdom. It also is currently unknown how effective the process is or what the potential long-term consequences of this manipulation are on the person in question and any future offspring. It also could lead to some murky ethical questions regarding parental rights since there are technically three biological parents. There also is the slippery slope argument of, if we start here, where does it go next? Does it turn into designer babies, which is a risk of PGD, or is it an exciting, life-changing process for parents who wish to have biological children that are healthy enough to survive childhood? There are also questions about what the rights of the child are, if any. The child obviously cannot consent to this process when they are single cells derived from consenting adults.

Summary

In this chapter, we examined some big ethical, moral, and legal dilemmas surrounding genetic technologies in today's society. We spent the bulk of the chapter talking about advances in genetic sequencing and the ethical quandaries that arise from the advent of fast, cheap genetic sequencing. We built on our understanding of sequencing by looking at various genetic modifications (GM). We discussed the role of GM organisms in our society, including the controversy around consumption of GM plants. In closing, we briefly covered several provocative genetic technologies that provide exciting new opportunities for society but are countered with several questions to consider as the technology advances. We will now switch gears and examine evolution (Chapter 8) and ecology (Chapter 9).

CHAPTER 8

The Diversity of Life: A Foray into Seashells

ONE OF MY favorite things to do is to walk in the surf along a beach. I could spend hours doing so (and have many times). I love the sensory experience of the beach ... the feel of the sand and the inward and outward movement of the water, the smell and feel of the breeze, the sight of the water, and the thrill of picking up seashells and identifying the wide variety of life that lives in the transition area between land and sea. I get a huge thrill out of looking for and identifying seashells.

In spring 2019, we stayed in an AirBnB along Florida's treasure coast with a private beach. Not only did a loggerhead sea turtle visit us to lay her eggs in front of our cottage (Figure 8), it was the best shelling I have ever experienced on any beach prior.

Figure 8. Loggerhead sea turtle tracks in Vero Beach, Florida.

It was my first time picking up channeled whelk, olive shells, shark eyes, jingle shells, and slipper shells. Of course, I found plenty of shells I had seen before—scallops, arks, and clams. Each time I went shelling, I would pick up all the interesting shells and take them back, lay them on the kitchen table, and identify them. I also would carry my camera to take pictures of critters that are not so easily brought back to the cottage for identification. Crabs, shorebirds, insects, plants, reptiles, anything living on the beach that I could get a picture of, were all fair game.

The diversity of life in such a specific area or habitat is astounding. If you extend that to the wide diversity of habitats around the planet, it is very humbling to think about all of the various life forms with which we share the earth. It is then not surprising to recognize that biologists are still discovering new life on the planet. Where and how did so many different life forms come to be? Why is biodiversity important? How do we experience biodiversity in our daily lives? These are just a few questions we will be exploring in this chapter.

THE "E" WORD.

Let's start at the beginning: how is it that there is so much biodiversity on the earth? It's due to evolution, the sometimes dreaded "E" word. I realize that culturally there can be contention with evolution, especially since some believe that evolution can conflict with dearly held religious beliefs. I've found working as a Sunday school teacher or small group leader in religious settings that explaining what evolution is and isn't can help alleviate concerns that evolution is in conflict with religious belief ... so please bear with me and keep reading.

Biological evolution is not the same as Pokémon evolution. Pokémon, or "pocket monsters," are a fictional group of organisms that gamers can collect in a virtual imaginary world. The Pokémon universe includes video games, movies, and a trading card game. Individual Pokémon evolve to become more sophisticated Pokémon, such as Bulbasaur evolving to Ivysaur and eventually to Venusaur. However, this isn't how biological evolution works. For one, evolution does not happen to individuals, but at the level of the population (or all individuals).

Evolution at the fundamental level is *descent* with *modification*. Descent refers to reproduction or the passing on of genes. Modification means changes to those genes. So descent with modification means that these modifications or changes can be passed down to future generations (because they are heritable). As we talked about in Chapter 6, our DNA can change over time via mutations, a completely natural process. What changes when or where is fairly random. There is a huge advantage of this process because it maintains genetic diversity (we'll come back to this later in the chapter) and can allow for new phenotypes, which are visible traits.

For example, blind cave animals (Figure 9) often possess a genetic mutation that led to the loss of eyes. Now there is a new phenotype: fish (or other organism) with no eyes. The benefit of these new phenotypes can confer some

Figure 9. Blind cave fish at the Denver Zoo.

kind of advantage that leads to improved reproduction (more babies!) and allows for the passing on of genes. Going back to our blind cave animals, in a dark environment there is no need for eyes. Therefore, animals that don't expend resources on making eyes can instead use those resources on other senses (smell, touch, and hearing) to be able to find food or mates. The lack of eyes is then selected for because animals without eyes are able to reproduce and pass on their genes better than those that do have eyes. New phenotypes are not necessarily a good thing though, and a phenotype that impairs the ability to pass on genes may not exist for very long. For example, if the phenotype makes it harder to find a mate, then it is less likely that the genes will be successfully passed on. The key again comes to a change (modification) that results in a more fit organism that is better able to pass on their genes (descent).

Another example of evolution that you may be familiar with in your daily life is the evolution of "super" bacteria. Have you seen the signs in the doctor's office about the use of antibiotics, stating that they aren't for colds (which are caused by a virus)? Or have you seen the sticker on a prescription of antibiotics that says, "Be sure to finish the entire prescription even if you feel better"? The reason for this signage is to help minimize the expansion of antibiotic-resistant bacteria. Overuse and inappropriate use of antibiotics drive the existence of antibiotic-resistant bacteria. Bacteria have evolved to have a gene that confers resistance to antibiotics. Taking an antibiotic selectively kills all bacteria that is not resistant to the medicine. However, antibiotic-resistant bacteria will survive and continue to grow. This is how bacteria strains such as Methicillin-resistant Staphylococcus aureus (MRSA) came about.

Not surprisingly given the number of sick people taking antibiotics and the use of antimicrobial cleaners, the first MRSA infections were reported in hospitals and assisted living facilities.

Bacteria with the gene resulting in resistance to antibiotic treatment have an advantage over bacteria that does not. Natural selection refers to organisms with some kind of advantage being more likely to pass on their genes. Over many generations, species can begin to look very different. Changes are accumulated as populations are affected by different external and internal factors. This is why natural selection is often discussed in the context of evolution; natural selection leads to evolution. The changes in DNA can be random, but the actual process of evolution is not random, because DNA mutations leading to favorable traits will stick around and be passed onto the next generation. This eventually can lead to new species as well (we'll come back to speciation later in this chapter). Natural selection is not the same thing as selective breeding. Selective breeding is a manmade method of deliberately breeding certain traits into an organism. For example, selectively breeding canines can result in desirable traits, such as herding, guarding and docility. Ultimately, domesticated dogs became very different from their wolf ancestors.

Sexual selection is an interesting phenomenon where evolution is driven based on female mate preference. The historical record tells us that Charles Darwin, who published *On the Origin of Species* and proposed natural selection formally for the first time (he wasn't the only one to think this was possible; Alfred Wallace presented a similar theory around the same time), had a real problem with sexual selection, particularly with peacocks. Aside from sexual selection, it makes zero evolutionary sense for male peacocks to have such fancy tail feathers. A big fancy tail seems more like a liability when trying to avoid being eaten by a predator and a waste of resources. This is a nice example of how culture has influenced the process of science, because when Darwin was writing up his works, he and other Victorians of the time couldn't comprehend that something so biologically important as evolution could really be due to something like *female* choice.

Studies with peacocks indicate that females preferentially mate with males that have elaborate tails and lots of eye spots. These elaborate tails are an outward show of healthy, fit males. Sexual selection is responsible for a wide variety of interesting phenomena in the animal kingdom. This includes the unique behaviors of the bower bird (including construction of elaborate bowers to woo females), the dancing behaviors and the gorgeous feathers found on birds of paradise, and the fighting behaviors present in many large mammals like elk and big horn sheep. These traits were selected for by choosey females as evidence of "fit" males worth reproducing with. Since in most organisms the female invests more in the offspring, it makes sense that the female would also be choosier about her mate.

Since evolution happens over generations, it can be difficult to clearly see it occurring, but that also depends on the generation time. For example, bacteria have short generations, and evolution can be seen in hours. Galapagos finches have a slightly longer generation time, and evolution can be seen over years. Humans have a very long generation time and it takes centuries to see differences. However, based on evolutionary theory, we would predict the existence of transitional animals, which are animals with characteristics of a previous form (like fins on a fish) and its new, yet future form (like limbs on a fish). One of my favorite memories from job searching a few years ago was visiting the lab of Dr. Jonathan Weinbaum at Southern Connecticut State University. He is a paleobiologist interested in evolution and dinosaurs. For a former nerdy kid, now nerdy adult, seeing *real dinosaur fossils* and talking to someone who digs them up for a living was extremely exciting. During the tour of his laboratory space, he showed me transitional fossils of dinosaur-like crocodiles called Archosaurians chronologically lined up in a row in his lab, so his visitors could view how the skull shape of the organism changed over time.

Is fossil evidence perfect? Nope. Conditions have to be perfect for fossils to form. It is estimated that the fossil record only represents about 0.1-1 percent of all extinct life. The issue with finding transitional fossils is just that: you have to find them. It makes sense that they would exist and they do (some famous examples include *Tiktaalik*, the fish with elbows, and *Ichthyostega*, or "fishfoot," a transitional fish with legs). Just because certain transitional fossils do not exist because the conditions were not right for fossilization to occur or they have not been found yet does not mean they did not exist. One of the very first lessons I learned about science in graduate school was that you cannot prove a negative. Just because you do not have something, or received negative test results, does not mean something does not exist or that there was nothing to find; it only means it is time to ask a different question, or the same question but test it in a different manner, or to keep looking.

Within the human population, you might observe a diversity of traits, like the fact that some humans are lucky enough to not have their wisdom teeth, also known as their third set of molars. The removal of my four impacted wisdom teeth while a teenager is an experience I'm thankful to not need to repeat! As anyone who has had these teeth removed can attest, it is not a fun process, and the telltale bruises and associated soreness and swelling take time to disappear. Some have theorized that the reason wisdom teeth become impacted (meaning, there is no room for them in the mouth and that instead, they will rest below the second molars potentially causing issues with infections) is because there was no longer an evolutionary advantage to having a large jaw with three sets of molars. The ability to cook our food makes it easier to chew and negated our need to have extra molars and larger, more powerful jaws. Since there was no longer an advantage for this, there was no selection perpetuating it. As time went on, some humans had a genetic mutation that resulted in a lack of wisdom teeth. It can be deduced that this is favorable, because if you do not have wisdom teeth in the first place, they cannot get impacted and lead to potential oral hygiene problems like infection. A lack of infection in the mouth makes it easier to consume food and likely appeals to the opposite sex, therefore leading to an increased likelihood of passing on genes (including genes associated with a lack of wisdom teeth) to the next generation. Without medical intervention, over time wisdom teeth could be lost altogether.

Humans carry a few traits left over from evolutionary time other than wisdom teeth. This includes our tail bones and tendency to get goosebumps, or the pilomotor reflex. We obviously no longer have tails, just the vestiges of tails. A vestige according to the Oxford dictionary is "a trace of something that is disappearing or no longer exists." Applying this definition to a vestigial organ or structure, it refers to an anatomical feature that harkens back to a more ancient structure—like our tailbones. Some species of snakes have vestiges of legs. Blind cave fish (Figure 9) have vestiges of eyes. At some point, the ancestor of the blind cave fish did have eyes, but a group of fish trapped in a cave eventually lost their eyes because they were not necessary in a dark space, creating a new eyeless species of fish.

Another interesting phenomenon demonstrating our relatedness to other species lies at the genetic level. Genetic evidence can provide more information about our family trees. For example, instead of grouping organisms based on shared traits, we now can examine how related different types of organisms are at the genetic level. There also is a sub-discipline of biology called evolutionary developmental biology, or "evo-devo." Evo-devo is based on the observation that embryonic development is similar across different species. For example, a set of genes called Pax genes controls eye development in humans, mice, fish, and flies. This is particularly shocking to consider since flies have compound eyes, which are very different from the binocular eyes found in humans, mice, and fish ... and yet the same genes drive their development. Without Pax genes, eyes fail to develop properly (if at all) in each of these species.

Evolution is fluid. There is no "perfect" organism. We see species as they are because they are the best form for the current environment. But there are certain evolutionary tradeoffs. One area of evolution that has received significant study over time is the relationship among the upright gait of humans, the size of our brains (particularly the neocortex, or front part used for thinking), and our pelvis size. The ability to walk upright and the development of a larger, more sophisticated brain were both distinct evolutionary advantages. However, walking upright requires a narrower pelvis.[14] A human baby still has to pass through that narrow pelvis to enter the world. What compounds this problem is that not only are we trying to fit babies through a narrower opening, over evolutionary time the brain and head that needed to fit through that pelvis also enlarged. So we are trying to fit a baby with a bigger head through a smaller opening. As I can personally attest to, no wonder giving birth hurts! Thankfully, I was able to birth my son without him becoming stuck (which can happen and requires emergency medical intervention). Although flexible plates in the skull (this is why babies delivered vaginally will often have a cone-shaped head and why babies have fontanelles, or soft spots) has helped the process, childbirth is still obviously an ordeal for the mother. But because walking upright and having a large brain are still more likely to lead to more reproduction than an easier time of giving birth, these traits persist.

A quick note on religion and philosophy before moving on. I bring this up, since, as I mentioned at the beginning of this section, the "science vs. religion" dichotomy is somewhat common in the United States and can be a major barrier to engaging with biology and science. Evolution is about descent with modification, and the study of evolution, like the study of science, cannot make any claim (to either the absence or presence) of the existence of any deity. Like I mentioned previously, just because we currently can't use existing science to generate evidence to support or refute the existence of a deity does not mean that evidence may or may

---

14 Going back to Chapter 2 and the fact that science knowledge can change in light of new evidence, there is some recent evidence to suggest that this may not be entirely true. See: Betti, L., & Manica, A. (2018). Human variation in the shape of the birth canal is significant and geographically structured. *Proceedings of the Royal Society B, 285*(1889), 20181807.

not be generated in the future. Individuals of faith might go so far as to tell you it isn't necessary to have that evidence to adhere to a religious practice because faith in a greater power implies having confidence in something without evidence or proof. "To create" as used in the Book of Genesis,[15] the first book of the Hebrew Bible and the Old Testament, has been suggested to have a different meaning to the original writers. Rather than the physical act of creating something (which is how our modern society uses "create"), "to create" instead meant to give purpose. For example, the table I am currently writing on can have the purpose as a desk for my computer or for an individual to eat lunch. Said otherwise, when God "created," s/he/they assigned purpose to humanity rather than creating it into being.

We see challenges in cultural interpretation in our society today, both between different cultures (for example, although Americans may talk about their fanny packs, "fanny" is slang for vagina in the United Kingdom) and generations (for example, having a davenport in your living room instead of a couch). This problem is compounded when we are putting thousands of years between the statement and the interpretation. This literal reading of Genesis, while attempting to be respectful of translating the culture and language of the original work, suggests a view of God's creation that does not conflict with evolutionary theory. Another possible way of thinking about evolution is that since evolution is not static and there is never a perfect final organism, this fits in with theological ideas about God actively creating in the world. This is common in mainstream Christian churches. For example, the United

---

[15] For a full treatment of this idea, I highly recommend John Walton's *The Lost World of Genesis One*.

Church of Canada creed says "God has created and is creating" in our world. Bottom line, there are many theological ways of thinking about what "creating" means that also are compatible with what we understand about biological evolution.

I am firmly in the camp that science and religious practice can exist in harmony with one another, because they can be thought of as two different domains of inquiry used for explaining the world around us and that each serves a distinct purpose in society today. One of my former students once said that science answers questions of how and what, whereas religion answers questions about why we are here and how to live. From my conversations with many scientists and people of faith, there seems to be more in common than differences between the two, since both fundamentally seek answers and require a discomfort from not knowing everything.

Evolution Produces New Species.

Eventually, genetic changes can accumulate until a completely new species is formed, especially in cases where one species may become separate from the rest of the population. Speciation is the term used to describe how new species are generated through evolutionary processes. Extinction, or the loss of a species, is the opposite. Defining a species can be challenging for reasons I won't go into here, but the generally accepted definition of a species is two organisms that can interbreed and produce fertile offspring. Over time, the ability to interbreed can be lost as what was once one species gets more different over time or diverges. Divergence can be caused by a variety of factors. For example, the organisms can have different areas where they live or different behaviors that prevent intermixing. In allopatric speciation, two species form after some kind of barrier (such as a mountain range or body of water) separates the two groups. What about creatures like a mule? A mule is the offspring of a donkey and a horse. However, a mule is not a separate species because it is sterile, meaning it cannot reproduce. So, according to our working definition of a species, since it cannot produce viable offspring, it is not a separate species, merely a hybrid of a donkey and a horse.

One of my favorite examples of speciation occurred with mosquitoes living in London during the Blitz in World War II. Since people were routinely staying overnight in the London Underground (London's famous subway system) to stay safe from bombing, mosquitoes that enjoy the taste of human blood followed. Eventually, the mosquitoes living underground diverged from the above-ground mosquitoes, creating a new subspecies of mosquitoes that are only found in the London Underground. This new subspecies of mosquitos have different traits, such as mating behaviors that are more reflective of a life underground. Mosquitos living above ground will swarm together to mate, but underground where it is less likely to find other mosquitos, a male and a female will pair off on their own rather than joining a swarm. This is a particularly interesting example, because the new subspecies of mosquitoes formed relatively quickly, after only about 50 years.

"WHAT'S THIS? WHAT'S THIS?"

In *The Nightmare Before Christmas*, a stop-motion animation film from the early 1990s, the main character, Jack Skellington, manages to leave Halloween Town and go to Christmas Town. Upon arriving in Christmas Town, he breaks into the song, "What's

This?", as he explores all of the novel aspects of Christmas Town. I sometimes feel like Jack Skellington whenever I'm in nature. Anytime my husband and I visit a new park or area of the world, he rolls his eyes as I purchase yet another field guide specific to the area we are visiting. My favorite field guide, not surprising given my descriptions of shelling from the opening of this chapter, is a beachcomber's guide to Florida's beaches.[16] My copy is much abused and filled with notes and dates of when and where I've observed various organisms. Formally, the branch of biology associated with the categorizing, naming, and identification of various organisms and how they are evolutionarily related to one another is called taxonomy. Perhaps you remember learning different taxonomic ranks (domain, kingdom, phyla, class, order, family, genus, and species) in school. Historically, notes such as mine have been useful to biologists for understanding evolution and ecology. It's not because there is anything special in my notes, but they do reveal what organisms were present in a particular place and time. Although I have no idea if my observation of a speckled crab in the surf in Vero Beach in May 2016 is really that critical to know, at the very least I was excited to see it and record it in my book. I like the biographical recording of all of my trips to the beach, with a nod to the fact that my scribbling may be relevant and useful to a biologist someday.

You don't need to travel to exotic places to identify and name organisms. In fact, you don't need to go too far at all. As much as I dislike my eight-legged roommates that have a tendency to invite themselves inside my house during the fall, I typically

---

16 *Florida's Living Beaches: A Guide for the Curious Beachcomber,* by Blair and Dawn Witherington.

observe more than a half-dozen different species of spiders over a few weeks during that season. I also have a bird feeder and finch sock in my backyard, and everything hangs conveniently at the window near the desk where I typically sit and write so I can watch the birds come and go. I regularly see a wide variety of birds over the course of my day. I've become old news to some of them, and they don't startle away when they see me working. The best part of living along a major migration route is that I also see seasonal visitors, which is exciting when I observe a bird for the first time. Just yesterday we had a flicker at the feeder. I keep a copy of *The Sibley Guide to Birds*,[17] a highly respected bird identification book, nearby to identify our avian visitors.

Next time you go outside, ask yourself about what you see. Even if you don't know the names, how many different trees or different birds can you find? Have you ever been outside and seen something new for the first time and wondered what it was? Identifying and appreciating the living world around us is one of the many ways we experience biology on a daily basis. It is an accessible and fun way to interact with nature. More than that, I would argue that it is good to take a minute and consider our place in the world and who our non-human neighbors are. I will talk more about this idea and the interconnectedness of life in the next chapter.

Biodiversity: What it Means and Why We Need It.

Although there is plenty of biodiversity to explore and experience just in our own backyard, there are certain hotspots around the world where there is a particular high diversity. Biodiversity hotspots are considered to be regions of the earth where there is

---

17 *The Sibley Field Guide to Birds*, written and illustrated by David Allen Sibley.

both a higher than average number of unique species, many of which can be found nowhere else on Earth, and species that face potential destruction from human activities, such as habitat destruction. Therefore, hotspots are both extremely unique because they are irreplaceable and also may be in danger of permanent loss. For example, let's consider the Amazon Rainforest in South America. The Amazon Rainforest has such incredible biodiversity that 1 in 10 *known species* on the *entire planet* are found in this single area, including an impressive range of insects, amphibians, birds, fish, mammals, and plants. In addition to being home to approximately 10 percent of all of the world's species, the plants within the Amazon Rainforest are responsible for pulling a substantial amount of carbon dioxide out of the atmosphere.

The problem is that the Amazon Rainforest is threatened, mainly due to human activities. Trees are cut down at an alarming rate to support farming and ranching practices. Since soil quality is surprisingly poor in this region, it is necessary to cut down more trees and plant in a different area every few years. This is why carnivorous plants like pitcher plants and Venus fly traps are found in the rainforest; the soil does not provide all the nutrients they need, so they instead turn to catching and eating insects. Loss of habitat then negatively influences the multitude of other species living there, and the result is a loss of biodiversity.

So why do we care? Some would take a moral stance and say that we should essentially be cleaning up our own messes and be better stewards of our natural resources. Since plant species in the rainforest are removing so much carbon dioxide from the atmosphere, their loss means that one of the greatest natural sinks, or storage, for carbon dioxide is gone. Burning plants releases all of the stored carbon dioxide back into the atmosphere increasing the problem. As we discussed in earlier chapters, carbon dioxide is a greenhouse gas, and so the more that is going into the atmosphere and the less that is being pulled out, the more we will observe climate change. Again, there is both a moral argument and a fiscal argument for caring since projections suggest that climate change is going to be a very expensive problem for humans to address.

Another financial argument is that maintaining biodiversity is a fiscally sound proposition. Ecotourism can be a financial boon to small economies. Having many unique and beautiful animals in a location is a draw for tourists. Drug discovery is another. So-called "medicinal bioprospecting" is the process of looking for naturally occurring compounds that can be utilized clinically to treat diseases. Much like we talked about in Chapter 3, this is not a novel idea. People have been using salicylic acid, which is found in willow bark, to treat headaches for centuries. One of the most commonly used chemotherapeutics (Taxols, known clinically under names like "Docetaxel" and "Paclitaxel") also comes from tree bark, this time from the Pacific yew tree. Perhaps you've experimented with pharmacologically active compounds found in the natural world, such as compounds in magic (hallucinogenic) mushrooms.

What is still left for us to discover, and how could those discoveries help us? The rainforest can recover … if it is allowed to while it still can. Feel the need to help the rainforest? There are relatively simple consumer decisions that you can make to help protect the rainforest. For one, avoid buying products that contain palm oil. Palm oil is in lots of common consumer goods from food to cosmetics. Cutting down rainforests to plant oil palms is a major driver of deforestation. Palm oil is both bad for your health (it is high in saturated fat) and bad for the environment. You can also buy responsibly sourced products. For example, look for the green frog emblem of the Rainforest Alliance on products like tea, coffee, and chocolate.

Biodiversity isn't just about the number of different species in an area, or how many of a certain kind of species are in an area, it's the genetic diversity, too. As mentioned previously in this book, genetic diversity is a good thing. In the event of a major conflict or threat, increased diversity will lead to an increased likelihood of survival. This is also why it's better for you to reproduce with someone who is genetically dissimilar to you. The more diversity the heathier the offspring.

Genetic diversity also is important when considering conservation efforts. Sometimes there is not enough genetic diversity left to save a species. The members that remain are forced to inbreed, and the resulting offspring are less healthy. Species Survival Plans (SSPs) are a formalized method to strategically breed endangered animals to help avoid this problem. Next time you go to the zoo, see if there is any signage around habitats housing endangered species to see if the zoo is participating in an SSP (Figure 10).

Figure 10. Species Survival Plan signage at the Denver Zoo.

The reason animals are often moved around the country, or even the world, is to facilitate breeding as part of an SSP. This isn't always the case, though, since animals in zoos can come in from illegal pet trade, and without knowing about the genetic history of the animal, it is typically considered better to sterilize the animals rather than allow them to breed.

Summary

Evolution is the patterns we see over time, and natural selection is the biological process that is responsible for speciation, or the generation of new species, and explains the wide diversity that is on Earth. Evolution occurs because random genetic mutations lead to changes in our DNA that are heritable, meaning genes can be passed down to our offspring. Said otherwise, evolution, or more specifically natural selection, is descent with modification. We see new phenotypes, or visible traits, appear because they provide some kind of advantage that eventually lead to increased reproduction. Evolution is an ongoing process, and there is never a "perfect" final result. Instead, we see organisms that are the best genetic fit for a certain place and time. Evolution is responsible for the incredible biodiversity on Earth that surrounds us. We can witness it in our own backyards or in one of the biodiversity hotspots around the planet. Biodiversity is a significant asset to humankind, and the loss of biodiversity has the potential to significantly negatively impact many species, including us. Biodiversity can be recovered, provided efforts are made in time. All life on the earth is connected to one another. Ecology is the study of the interconnectedness of life on the earth and the subject of the next chapter.

CHAPTER 9

THE INTERCONNECTEDNESS OF LIFE AND ISSUES OF CONSERVATION

Figure 11. A portion of my cloth diaper "stash."

OKAY, I'LL ADMIT it. I am a cloth-diaper enthusiast (Figure 11). Initially, it started out simple enough. Standard disposable diapers hang around in landfills for a long time. Every single one of the thousands of disposable diapers my parents slapped on me *is still sitting in a landfill*. Lovely. If you were diapered with standard disposables, all of your diapers are still sitting somewhere, too. If we piled up all the used diapers in landfills, I wonder how tall it would be. Mount Dirty Diaper. Gross.

Using cloth diapers saves money, too! Well, unless you start collecting them like I do. But seriously, if you don't start collecting them, you can take care of all of your diapering needs for every kid and beyond, too. I did a quick informal poll of the online cloth diapering community as part of research for this book. About 17 percent of people who responded to my poll are using the same diapers their parents/in-laws used for them or their spouse. All the diapers you'll ever need you can purchase for a few hundred dollars depending on if you go with prefolds and covers ($100-$200) or fancy all-in-one diapers that function like a disposable diaper (~$500). It's even less if you buy used and then resell everything when you're finished. Believe it or not, there is a huge market for second-hand poop catchers.

But wait! Is cloth diapering actually more green? It takes resources to grow cotton, then weave it into cloth, then cut the cloth and sew it into diapers, and then ship said diapers to a store that then ships it to consumers. I also have to wash my diapers, and not surprisingly, I have to use more water and soap than I would with a typical load of laundry. Since most cotton production isn't in the United States either, there is at least one trip across an ocean for raw materials. This isn't a unique problem to cloth diapers either … the fashion industry is listed as one of the most polluting industries in the world because things get shipped around so much and the whims of fashion dictate new styles every few months. So clothing like shirts and pants ends up being resource-costly to produce … and then they go out of style before the end of their usable lifespan.

Really, why should I care anyway? In spite of my collection, I'm still paying less for diapers than if we were using disposables. I'm saving a buck and my son's fluff butt is cute. We have zero plastic bags in the house and use wet bags[18] for everything. Plus, my wet

---

18 Wet bags are water-resistant bags initially intended for storage of dirty cloth diapers before washing. Sometimes they have two pockets, one for storing clean diapers and the other for dirty diapers. They are also useful for a wide variety of non-diapering things from toy storage to dirty gym clothes.

swimsuit doesn't leak through a plastic shopping bag and make a mess. I have a tidier house because I don't have a bag of plastic bags taking up space. So, why does anything else matter?

As I mentioned in the close of the last chapter, life is very interconnected, and seemingly small decisions, such as deciding to purchase a reusable instead of a disposable product, can have a large impact. Ecology, the subject of this chapter, is the study of interactions or, more specifically, the relationships between organisms (including us) and the environment. Disturbing those interactions and relationships can have far-reaching consequences and lead to several conundrums about how to balance modern society with an ecologically grounded mindset. This chapter is split into two parts. First, I'll introduce the study of ecology. Then the rest of the chapter will focus on one of the most critical applications of ecology to today's society: conservation issues and the consumer decisions around going "green."

What is Ecology?

Ecology is the study of interactions, and these interactions occur at multiple different levels or scales. At the largest level, there is the biosphere, or all parts of the earth where life exists. Then we have the ecosystem, which describes how both living and non-living things in a particular area interact. For example, in the alpine region of the Rocky Mountains of Colorado, the plant and animal species (living) are well suited for the cold, dry air (nonliving). The next level, communities, only concerns the different species living in a particular area. Population is the total number of any given organisms in an area, and, finally at the smallest level, we have the individual organism.

We hear about population fairly regularly in the news. Perhaps you've heard something about the growing human population, or the numbers of a certain pest species, or that an endangered species is bouncing back (or not). If you happen to hunt or fish, you're familiar with getting hunting or fishing licenses that are designed to help control populations. For example, hunting too many or too few of a particular animal (let's say deer) during hunting season could have serious repercussions on the population in the future and impact everything from future hunting seasons to the number of deer-car collisions.

How many of a particular organism exists at any time and place is the result of additions (births or migration in) and subtractions (deaths or migration out). Populations of mice tend to increase more rapidly than populations of humans because they typically reproduce more, with more babies born each time, and become sexually active much sooner. Humans typically only reproduce a few times and have a longer gestation period. In the United States, it is estimated that households have approximately two children, which translates to approximately two reproductive episodes with a litter size of one. It is not nearly as common for humans as it is for mice to have a litter (twins, triplets, quadruplets), let alone a litter every few weeks, and consequently our population grows slower.

Populations are (generally) limited by resources as well. These resources include everything from food to places to live (habitat). Have you heard of an invasive species before? An invasive species is an organism (either plant or animal) that is introduced somewhere new whose population then booms in the new location. Why does the population boom? Well, it could be because natural predators are not present in the location, or there is a new source of food, or a bit of both. If you like spending time outdoors, you may be aware of efforts to try to stop the spread of various detrimental invasive species. Have you seen signs cautioning you to not move firewood? These are efforts to stop the movement of the emerald ash borer, which is systematically killing all of the ash trees in North America. It is native to Asia and thought to have been accidentally carried to North America in wood-packing materials. Avoiding carrying pests and diseases between continents is why there are such strict rules regarding what you can and cannot bring with you during international travel or to places like the Hawaiian Islands. You can identify trees killed by the ash borer by the characteristic serpentine patterns on the trunks of the dead trees (Figure 12).

Figure 12. Ash tree killed by the emerald ash borer. The serpentine marks are a hallmark of trees killed by the emerald ash borer.

How do we know if efforts to control invasive species or other animals or pests are working? We track the populations over time. If we want to control a population of nuisance animals, we can target either the birth or the death rate. For example, koalas are given birth control implants or are sterilized to control their population by decreasing the birth rate and consequently limiting the population. Increasing the death rate also results in a decrease in population. For example, hunting is used to attempt to control nutria, which are large rodents, in Louisiana as well as Burmese Pythons in Florida. Both of these species are invasive and cause major environmental damage.

As I mentioned at the opening of this chapter, ecology is about interactions. Although something like a large deer population resulting from a lack of apex predators (top predators like wolves or mountain lions) may seem innocuous (deer are cute and won't eat your dog), the out-of-control population has a substantial impact on the environment and on our wallets. Deer strip down young trees when they eat the leaves, which can ultimately result in the deaths of the trees. Without the trees, there is loss of habitat for other species. Those species not having a place to live then can go on to have other impacts on the environment. Disturbing something in an ecological system has the ripples-on-a-pond effect ... and we cannot always foresee just how extensive the consequences will be.

I am originally from Pennsylvania, which, according to estimates from State Farm, is one of the top states in the United States for deer-car collisions. A study by State Farm indicates that the average cost of damages in a deer-car collision is more than $4,000. Ouch. Guess why people are paying more for car insurance?

Although the decision by humans to cull apex predators like lions and wolves may have made logical sense from the prospective of protecting livestock, fiddling with the ecological balance ended up having a financial cost down the road. This brings us to the second part of the chapter, how do we, as humans, interact with ecological systems (for the better or worse), and how do we make biological sense out of our place in the world?

Questions of Conservation: Finding Balance.

When talking about the conservation of resources or animals, it is very easy to fall into "doom and gloom" patterns. This is a major issue with increasing public awareness regarding how much humans have disturbed the biosphere and how dire the situation has become. There is plenty of evidence suggesting that our current trajectory is not sustainable ... but, how do we get attitudes toward conservation to change? Not surprisingly, focusing on the "doom and gloom" has a tendency to turn people off to engaging with these issues, let alone spurring action to do something about them. We, as humans, are just as much a part of the ecosystem as any other organism. The difference is, we have a tendency to wreak more havoc than other organisms and throw things out of balance.

So, what do we do? How do we balance the needs of humans AND the other organisms with which we share the planet? It starts with the following realization: that we are interconnected with all life on Earth. With the growing human population, it is necessary to identify creative and sustainable ways of feeding and housing humans, while also acknowledging that we are not the only ones living here. Let's look at a few examples of ecological and conservation issues with an eye to creative solutions and lingering problems, including actionable decisions that you may come across in your daily life.

The Bushmeat Crisis: Save [insert endangered species here], or Feed Starving Humans?

It is really easy for me to sit in my house with a well-stocked pantry and short bike ride or walk to the grocery store and get incensed about people in poor countries eating endangered animals. Which is more important, conservation or feeding the hungry? Is this a #firstworldproblem since I ate today and have plenty of non-endangered animal protein sources available to me? For many people, eating endangered animals is the only reliable source of protein. The catch, however, is that eating animals that already have dwindling numbers is not sustainable.

In Africa, the bush is what we in the United States would call the woods or forest. Bushmeat is wild game. For many indigenous peoples of Africa, eating bushmeat was a key and essential part of their diet. However, with international trade and commerce and the advent of logging roads into the bush, commercial, illegal, and unsustainable hunting practices became easier and the result was the decimation of animal populations that were previously too remote to be threatened. The "bushmeat crisis" refers to the major loss of biodiversity resulting from these new and unsustainable hunting practices.

Oh, but it gets worse and starts to hit a little closer to home. The consumption of bushmeat is also linked to the emergence of several deadly diseases in the human population. Research suggests that the Human Immunodeficiency Virus (HIV), which is the virus that eventually leads to Acquired Immunodeficiency Syndrome (AIDS), is thought to have passed to the human population after contact with bushmeat. The largest Ebola (hemorrhagic fever) outbreak ever reported occurred in 2014 (which included a few people in the United States getting sick) and was hypothesized to result from humans either eating or having direct contact with fruit bats.

To recap, bushmeat consumption is not a good thing because it tends to be unsustainable, and laws designed to protect animals are not enforced, so biodiversity is lost. As we talked about in the last chapter, biodiversity is really important, and as discussed earlier in this chapter, everything is interconnected, so the loss of one species can have long lasting complications, both for other species and the planet. Bushmeat also can be a reservoir for dangerous diseases like Ebola, and with international trade and travel, diseases can travel to different parts of the world.

However, bushmeat is a traditional source of protein in Africa and, for many people, the only reliable source of protein. Again, we get back to the issue of balance: how do we minimize the spread of deadly diseases, preserve biodiversity, and feed starving children? One proposed solution involves importing chickens and teaching villagers how to raise and process chickens. Then the chicken, rather than some endangered primate, becomes the source of protein. Another possible solution? If you happen to be traveling internationally and see the opportunity to eat bushmeat, don't do it. Sure, if you have an adventurous pallet, the opportunity to eat chimpanzee might be appealing, but really, the odds are you are contributing to the problem. Money talks. If there is no market for exporting illegally harvested animals, it is going to decrease. Speaking of money, let's move on to a major consumer decision in the United States today: the decision to reuse or not.

Is my Reusable Grocery Bag Actually Greener?

I opened this chapter with the factors leading to my decision to cloth diaper and how, although it seems like it should be the greener, more eco-savvy option, there are certainly tradeoffs involved. When it comes to making ecologically sound decisions, it is important to look at the shades of gray and determine which tradeoffs make sense and which do not. Some eco-friendly decisions make sense and do not have as significant of tradeoffs. For example, I ride my bike regularly for running errands. I'm also a regular customer of Goodwill, thrift stores, Craigslist, and consignment sales. I could buy new if I wanted to, but I'm (a) cheap and (b) aware that there are finite resources available on Earth and would rather buy something already produced than spur the manufacture of more goods.

Not all consumer decisions are this easy though. What about grocery bags? Plastic bags are cheap and easy to produce. They also are suffocation hazards and take approximately 1,000 years to degrade in a landfill. They shed microplastics (microscopic plastics) that we then go on to ingest in our food. However, they take fewer resources to make than a reusable bag. Then if you forget your reusable bag, you buy another ... then you repeat that at the store the next week. Before long, you've got a large stack of reusable bags at home. Reusable bags tend not to be recyclable either and once they finally reach the end of their usable lifespan, they, too, end up in the trash.

If we can recycle plastic bags (or any plastic packaging), isn't that the better option? The issue with recycling is that there needs to be a market for the raw recycled materials. Otherwise, the plastic bottle or bag ends up in the landfill. It is not always financially feasible to recycle plastic … so it ends up being cheaper to make more and dump the used stuff in a landfill or the ocean. The same problem occurs with glass and some metal waste. Although glass or metals like aluminum are endlessly recyclable, consumer recycling doesn't make sense from a financial standpoint, so it doesn't always happen. Some companies such as RePlay and Green Toys are creating a market for raw recycled materials. Both companies take used milk jugs and turn them into new products. RePlay makes children's dinnerware, and Green Toys makes … well, toys (Figure 13).

Figure 13. Garbage given new life as a toy. My son's barn set made from recycled plastic.

The other factor in the decision to reuse is how much do we value landfill space? As part of research for this book, I was chatting with Stephanie Daniel, the owner of an online cloth diapering and environmentally friendly boutique. She pointed out that she could justify cloth diapers, in spite of all of the extra resources that go into their production, because landfill space is not a renewable resource. By piling up all of our garbage, we lose land, space, and habitat. Habitat not only for animals, but for ourselves as well. We have a growing human population that needs more food to eat and places to live; plus, we face the threat of lost living space due to rising oceans and ever growing piles of plastic waste that won't disappear anytime soon. We are still connected to and impacted by our garbage ... even if it is generally out of sight, out of mind.

Another issue on the financial piece is the cost of going green. Let's go back to the cloth diaper example again. All-in-one cloth diapers (essentially a cotton or hemp version of a disposable diaper covered in water-resistant fabric) are difficult to produce. It takes skilled workers significant time to make a diaper that will actually work. How do companies produce an ethically made diaper that is still going to be affordable to the average consumer? Tradeoffs again ... by purchasing fabric produced overseas. Textile manufacturing (as well as the manufacturing of many other products) has moved overseas. Even diaper brands like Thirsties that boast United States-based production are buying raw materials from overseas. Thirsties, however, does the remainder of its production (cutting and sewing) in Colorado and tries to offset resources used in diaper production by using sustainably derived energy to power their warehouse. Again, it's about the issue of tradeoffs. If a diaper was made out of raw materials grown and woven here in the United States, it would be too expensive for the average consumer. For example, during my chat with Daniel, she was telling me about a brand that produces children's tableware from bamboo in the most ecologically sustainable fashion possible. They opted not to carry the brand in the store due to cost ... a single plate is $25+.

Quality or quantity? High upfront cost? Or larger, less identifiable cost over time? If I'm going to buy new, I always go for quality. However, I also recognize that I'm in a place financially where I can buy a high-quality, environmentally friendly wooden toy or purchase furniture made from wood and not particle board that is going to warp and degrade faster over time. My hope is that our culture will shift from a focus on overproduction of cheap disposable items and buying moremoremore and move toward the production of affordable, high-quality goods. I drive a 13-year-old car. I adore it. We opted to spend a bit more money when we bought it to get a quality used car, and 140,000 miles later and after multiple cross-country trips, it remains a reliable car. I don't care that it's an "older" car—it works well for us, and the maintenance has never been excessive on it, which matters far more to me than keeping up with the Joneses.

As I mentioned before, money talks. Next time you are at the store ask, what are you spending your money on and why? Do you need to buy a "thing?" Are you "voting" with your money by purchasing cheap things that will end up in a landfill or being more selective? Is what you are buying from a sustainable source or subtracting from the finite resources on the planet? Is it made in the United States? Is it made out of a natural material that will degrade in a reasonable amount of time? Are you buying a material gift that could be substituted for an experience instead?

Composting and Organic Waste.

Speaking of degrading, where does organic waste go? Organic waste is anything that is carbon based and will break down over time. The ecosystem has a great way of recycling organic waste and turning it back into usable resources. Although this is a completely natural process, it requires certain conditions to work as my husband and I found out when we first started adventuring into composting, or allowing our organic waste to turn back into organic matter.

When I bought my first house, the previous owners left behind a composter. It was one of the round, black R2D2-esque structures and was sitting in the backyard, strategically placed behind a bush. Great! We could compost! I had seen my grandfather dutifully adding organic waste to his compost pile behind the garage for years—how hard could it be? I mean, composting is a natural biological process that happens on its own all the time, so how difficult could it be for us to compost? Weeelllllll, we still managed to mess it up. Rather than generate compost, all we managed to do was mummify some banana peels. Where did we go wrong? Things rot just fine on their own in a forest. Rot is a super important part of the ecosystem after all.

It turns out composting requires a certain set of ingredients … and we were missing some of the ingredients. We had organic waste (banana peels) … and we watered our compost, which is the second important ingredient. There was access to oxygen in the air, which is the third ingredient. What we were missing was soil with a healthy population of microorganisms, particularly aerobic bacteria. The bacteria (and other microscopic organisms like fungi and protozoa) are the "workers" and break down the organic waste into nutrient-dense compost. They "eat" the food waste, "breathe" the oxygen, and "drink" the water. They also produce carbon dioxide gas and heat during this process. This is why compost heaps get warm.

Have you heard of people turning and watering their compost heaps? This speeds up the process and keeps the bacteria and other microorganisms happy. If there is too much heat, the microorganisms will die off. When I found out about curbside composting (our organic waste gets picked up weekly, and in return we get compost back in the spring), I was really excited—it was an easy way to be earth-friendly. My husband asked, "But wait—if this stuff naturally breaks down, why are we paying extra per month to *not* send it to a landfill?"

Great question. Here's the answer. There are many different kinds of bacteria, but we're only going to focus on two: aerobic and anaerobic bacteria. When you think of aerobic bacteria, think about aerobic exercise—what do you do during an aerobics class? You breathe heavily because you need more oxygen. Aerobic bacteria are the same way—they need oxygen, too. Anaerobic—think "an" as in opposite; these bacteria are the opposite of aerobic bacteria and don't require oxygen. Remember Chapter 5 when we talked about energy? Aerobic bacteria require oxygen so they can do cellular respiration to make energy.

Landfills are anaerobic. We put our garbage in plastic bags, which block the flow of oxygen, and then as garbage piles up in the landfill, it also blocks airflow, basically suffocating the garbage and any resident aerobic bacteria. Which bacteria live in landfills instead? You guessed it—anaerobic bacteria. Anaerobic bacteria produce methane when they eat up our food waste, and methane is a very potent greenhouse gas. A greenhouse gas is a gas with an insulating effect in the atmosphere—think of it as slowly creating a thicker and thicker blanket of atmosphere around the earth. The thicker the blanket, the warmer the earth's surface will become. The Environmental Protection Agency states that landfills are the third largest producer of methane gas in the United States.

Hang on though—isn't carbon dioxide a greenhouse gas, too? Yes! But methane is about 20 to 25 times more potent a greenhouse gas than carbon dioxide. More potent, meaning it contributes more to the "blanket" warming up the earth. I definitely fall into the category of "too busy" to compost in my backyard, but I'm trying to do my part to be eco-friendly, too. So, no, I don't mind paying a few extra dollars a month for curbside composting. Am I going to change the world one 5-gallon bucket at a time? Not by myself—but if enough people start composting, not only is that less methane in the atmosphere, but it will save room in the landfills and ultimately what we spend in tax dollars. Plus, I won't need to buy any fertilizer next year for my yard or garden.

Ecological Crossroads and Hardin's Tragedy of the Commons.

There is a fantastic meme I've seen floating around the internet. It basically says we don't need a bunch of people going zero waste perfectly, but more people making an effort to decrease their waste. I'm not going to change the world by composting … but if you look at everyone in my area that uses the same composting service, we are talking about thousands of pounds of food scraps not ending up in the landfill. What are some tradeoffs that you can think of in your own life? Are there small things you can do to be more ecologically minded? Although the rapid loss of biodiversity, polluted environments, and environmental destruction seem insurmountable (all that doom and gloom), it is still possible to do good and fix things, and it starts easily enough with the decisions we make every day. What if you bought a refillable water bottle instead of a case of bottled water the next time you went to the store? What if you started taking paper instead of plastic bags at the store, or better yet made a commitment to bring your own bag? What if you shopped at consignment sales instead of always buying new? How much money will you personally save and how much benefit will the biosphere reap as a result?

There is an ecological principle called Hardin's Tragedy of the Commons.[19] In a shared-resource system (like an ecosystem or the biosphere ... or something like a shared kitchen area), if everyone acts in their own interest rather than for the common good, eventually the resources needed by all will be lost by all. Look at the example of the shared kitchen area. If everyone leaves a dirty dish in the sink, eventually the sink will be unusable because it is overflowing with dirty dishes. In the case of a fishery, if everyone takes a little bit of extra fish (it's just a little, right? it won't matter.) so they have added food security, the population of fish eventually collapses. If everyone had just taken exactly what they needed, the resource would have been preserved for everyone. Again, ecology is the study of interactions ... everything we do has a push-pull interaction with both the humans around us and the other organisms.

SUMMARY

In this chapter, we started with the basics of ecology and examined the interactions around us and closed with discussing some ecological conundrums we find ourselves in. Since we are part of an ecological system, our decisions impact other organisms around us. Sadly, many of the decisions made by humans, particularly in the last 200-ish years since the start of the Industrial Revolution,

---

19 Hardin, G. (1968). The tragedy of the commons. *Science, 162* (3859), 1243-1248.

have severely and negatively impacted other organisms on this planet, including us. In spite of dire statistics about loss of biodiversity and polluted ecosystems, there is still opportunity for things to rebound. Sure, you can focus your attention at a bigger scale and volunteer or advocate, but there are decisions we can make at home as well, ranging from our consumer decisions to turning the lights off when we leave the room. It's all about developing the habits, making the choices, and building an ecologically focused mindset. Speaking of the human mind, we are going to completely switch gears going into the next chapter and talk about a topic at the intersection of biology and psychology: human growth and development.

CHAPTER 10

Tiny Humans: An Expose at the Intersection of Psychology and Biology

CONSIDER FOR A moment that at one point you were a single cell. Less than a year later, you were a completely helpless infant dependent on the goodwill of adults around you to see to your well-being. Somehow over the last how many years, you've gone from a single cell to a baby that could live in the world; you learned to talk, walk, read, feed yourself, use the toilet, play a sport, drive a car, crochet, whatever. Maybe you've even reproduced and have tiny humans of your own. If you consider the entire process from start to finish, it is truly awe-inspiring that we all start with such humble beginnings and grow to fully functioning humans.

No matter what your journey through life was, you've done just that: you had an incredible journey from a single cell to an adult human being. The interesting thing about human development is if you want to take a course that covers conception to birth, you would look in a biology department for a developmental biology course. If you are interested in what happens on the journey from birth to adulthood, you actually need to stop by a psychology department and sign up for a human development course. In this chapter, I'm going to attempt to do both by starting with conception and ending with adulthood. In particular, since this is at its core a biology book, I'll point out areas where biology and psychology overlap.

In the Beginning ...

You were two single cells, an egg from Mom and a sperm from Dad. Typically, Mom only has one egg available for conception every month (guess what happens if the egg doesn't meet up with the sperm? A woman gets her period), and the egg is only viable for a short period of time. Sperm can live longer in the reproductive tract, but in spite of that, there are still only a few days a month when the egg and sperm can come in contact with one another. Ovulation, or the process of releasing an egg, typically only occurs once per month, and the egg is only viable for a few days. If sperm are already present prior to ovulation, or find their way immediately after ovulation, conception occurs. Conception happens when the egg and sperm fuse together to form an entirely new being ... you.

What happens next? Well, the zygote, or the initial cell that forms after conception, hangs out for awhile going through mitotic divisions (to make more identical cells) until it eventually finds its way to the uterus and implants. By this point, it's called a blastocyst. Embryonic stem cells come from blastocysts. One of the reason people object to their uses is because the blastocyst is on its way to becoming a baby, although at the time stem cells can be harvested, it is only a mass of stem cells.

Once the blastocyst implants in the wall of the uterus, Mom and the future baby have established a physical connection. Have you heard of an ectopic pregnancy before? This occurs when the zygote doesn't find its way to the uterus and implants somewhere else, causing a life-threatening medical emergency. Implantation can take anywhere from seven to 14 days. Some women say they can feel it, and others observe a small amount of bleeding at the time of implantation. Once implantation occurs, the real fun (and biological magic) begins. A positive pregnancy test usually follows a few days later. Even before that, astute women usually can tell they are pregnant because of the amazing changes that begin with the process of implantation.

In my personal experience, it was extremely obvious with my sore breasts and falling asleep anywhere all the time that I was pregnant days before I took a pregnancy test. Even more interesting, a few days before my period was due, I became extremely sick. I remember sitting at the doctor's office getting diagnosed both with bronchitis and strep throat and the doctor asking me if there was any chance I could be pregnant ... there are immunological changes that happen at the start of pregnancy to help the woman's body be able to host a foreign creature. So, the fact that I simultaneously came down with multiple illnesses at one time right around the time of implantation was a telltale sign I was pregnant.

So, what happens after implantation occurs? The blastocyst develops further and goes through a process called gastrulation, which begins the process of forming specific cell and tissue types. The placenta begins to form. The placenta is a fascinating structure because it is grown especially for each pregnancy, and then lost afterward. The placenta and the umbilical cord connect mother and child and facilitate the exchange of nutrients and wastes. The placenta also has a protective function. For example, there are enzymes in the placenta that break down stress hormones before they can reach the fetus and mitigate any possible damage to its development. We touched on epigenetics and prenatal programming briefly in Chapter 6 and will return to it again shortly. Around four weeks after conception, the blastocyst graduates to embryo status and has primitive organ structures, such as a rudimentary heart and brain. By 12 weeks post-conception, the embryo graduates to fetus status. The fetus stage lasts until birth and is when the rudimentary organ systems continue to develop and the fetus grows exponentially in size.

In Chapter 4, we discussed how important time and place are for proper biological processes. It is no different in prenatal development. There is a great quote by medical geneticist Veronica van Heyningen stating, "The amazing thing about mammalian development is not that it sometimes goes wrong, but that it ever succeeds." The right genes need to be turned on in the right place and at the right time to get a fully formed and functional organism. One interesting example is the Hox genes. Hox genes specify body placement, so your arms end up being in one place and your legs in another. The interesting thing about Hox genes is that the order they sit on DNA reflects the order of proper body formation. There was a famous experiment done by developmental biologists using fruit flies where they switched the Hox genes that specified legs for those that control antennae development. Guess what happened? The flies had legs growing out of their heads instead of antennae. Going back to Chapter 8 and evolution, Hox genes, like Pax genes, also are highly conserved across the animal kingdom.

It is not only the right time and place that are important, but the right amount, too. Sonic Hedgehog (abbreviated SHH) is a protein involved in pattern formation. SHH is expressed in a gradient during the development of specific structures, like the hands. If there is more (or less) SHH, different structures will form there. One classic example of proper pattern formation is getting the right number of digits. Polydactyl is a condition where an organism has too many digits, caused by defects in the amount of SHH present. If you've ever been to Key West, Florida, you may be familiar with the so-called Hemingway cats. These are cats that descended from a polydactyl cat given to Ernest Hemingway, who loved polydactyl cats, as a gift. And now there are dozens of cats with extra digits inhabiting Key West, many of which live in the Ernest Hemingway House. It isn't just cats either. Humans can have extra digits on their hands or feet, too.

PREGNANCY AND NO-NO LISTS.

We certainly can get developmental hiccups from problems with genes being expressed in the wrong time or place ... but the environment also plays a huge role in proper human development as well. The prenatal and, as we'll come back to in a minute, the postnatal environments are critically important for proper development.

Let's start with the prenatal period. Consider for a moment the things that are off limits to pregnant women. Well, there are teratogens. A teratogen is anything in the environment that causes damage to the developing fetus. This includes well-known teratogens, such as alcohol and cigarette smoke, and certain prescription medicines, such as isotretinoin, which is used to treat acne. Unfortunately, we are lacking good research on the safety of many prescription and over-the-counter drugs, which means, they, too, are off limits to pregnant women. Part of the reason we do not have data is because it is risky to conduct research with pregnant women, since it could permanently harm the developing fetus. So, women get by, and in situations where the risks of not taking a medication outweigh the risks of taking it, women and their fetuses are closely monitored. This is the same case with breastfeeding mothers. Since the majority of medicines will pass into breast milk at some level, many are not approved for breastfeeding women or state that they lack sufficient evidence to say whether or not they are truly safe.[20]

There also are limits on what women can eat while pregnant. I desperately missed unpasteurized soft ripened cheeses like blue cheese and gorgonzola during my pregnancy. Unpasteurized means the milk is not treated to remove potentially infectious microorganisms that can add flavor to the cheese but also can cause problems like listeria. Listeria infection while pregnant can lead to a miscarriage or stillbirth. Many varieties of seafood also are off limits due to mercury levels. We covered bioaccumulation in Chapter 5 and how toxins like mercury can build up in apex predators. Prenatal exposure to mercury can cause significant physical and intellectual problems.

What about epigenetics and prenatal programming? In Chapter 6, we talked about genetics and epigenetics and introduced prenatal programming. We are only beginning to scratch the surface of understanding the interaction between the environment, gene expression, and later health and wellness. There are thought to be sensitive periods during both prenatal and postnatal development. These sensitive periods can be used to refer to particular times

---

[20] Interested in helping out with this research? You can check out ongoing studies at the Infant Risk Center at Texas Tech University. Visit https://www.infantrisk.com/ for more information.

of development when the fetus or child is particularly susceptible to environmental influence, for the good or worse. For example, there may be periods during prenatal development where the child is more susceptible to environmental toxins, perhaps during the formation of certain organs. Sensitive periods also can be used to refer to periods where a child is more likely to pick up a certain skill, such as learning a second language.

Congratulations, You Have a Tiny Human! Now what?

In spite of the fact that humans are born relatively helpless compared to the rest of the animal kingdom, they do possess a surprising number of skills. For example, it is now becoming popular for healthy babies to be placed skin to skin with their mothers immediately after birth. Skin-to-skin contact helps babies (and Moms!) adjust and eases the babies' transition to the outside world. Skin to skin helps babies regulate their body temperatures and calms them so they cry less, plus it facilitates bonding and breastfeeding. All we did the first 24 hours of my son's life was sit skin to skin. If he wasn't perched on my chest, he was sitting on my husband's chest. Anyway, back to the immediate period after birth. Babies placed on Mom's stomach or chest can instinctively crawl to the breast and begin to breastfeed. Completely independently. It is thought that the reason a woman's nipples darken during pregnancy is to help the baby with their immature eyesight find a nipple and initiate breastfeeding. The nipple also smells like amniotic fluid, which the baby recognizes and can use for navigating.

The "breast crawl" is only one of many capacities held by newborns. There also are many reflexes, which are automatic responses to certain stimuli. This includes the rooting reflex, which helps the baby breastfeed, or the Moro, (startle reflex) when a startled baby flings out their arms when afraid. Sucking is a strongly ingrained reflex and facilitates breastfeeding. Since breastfeeding also serves emotional regulatory purposes, it makes sense that babies or young children will suck on a pacifier or thumb to calm themselves. Other than being interesting to see, why else do we care about reflexes? Well, pediatricians can test the reflexes and ascertain the health of a baby's nervous system. Inappropriate reflexes could be indicative of brain damage, and so testing them can help pediatricians determine if intervention is necessary.

Parents who joke about newborns doing nothing but sleeping, eating, and pooping aren't kidding. In the early weeks, there are lots of naps, snuggles, and crying. Crying is interesting because of the biological impact it can have on caregivers. Someone else's crying baby is annoying … your crying baby is more distressing than annoying. Interestingly, one of the best ways to help newborns adjust to life outside the womb (and cry less) is to recreate the experience of being in the womb. This again gets back to issues discussed earlier in this chapter as well as in Chapter 7. Human babies are born relatively immature, partly because if they stayed in Mom for longer, they would never fit out of her pelvis. So, recreating the womb-like environment through white noise or swaddling helps the baby adjust and tired parents sleep. It's a win-win for everyone.

Infancy and toddlerhood, or birth to two years, see a greater period of physical growth than any other time throughout the lifespan, although most of the growth is body fat and not muscle. The brain in particular grows incredibly fast, faster than any other organ in the body. The time when rapid brain development occurs is one of those sensitive periods I mentioned earlier. Without proper stimulation or loving relationships, the trajectory of brain development can be permanently altered. For example, the stress on children from living in poverty can undermine the child's ability to learn. Even from the newborn stage, babies are attracted to novel stimuli and can imitate adults' facial expressions. In the latter parts of the first year, babies begin to develop motor skills, including things like the pincer grasp when they can reach down and pick up something like a Cheerio and eat it. Language development goes from crying in the first months to cooing and babbling, and then eventually to saying first words. Interestingly, babies learn to understand language far before they are able to use it. So, yes, keep talking to babies! They understand more than you realize. "Baby talk," or as it is known in the developmental psychology literature, infant-directed speech, is actually good for language development.

Reading, particularly didactic reading (which is a more interactive form of reading that stimulates discussion, rather than a parent simply reading the book), even from the early days, is extremely important not only for language development, but also for socio-emotional development. Humans are emotional beings, and emotions begin to develop during the first few weeks of life, usually when the baby begins smiling back at caregivers. Later in the first year, infants start to do something called social referencing, where they check to see a caregiver's response to a situation and model their behaviors after the caregiver.

I have a funny anecdotal story about social referencing. We took our son to a local children's museum when he was around a year old and just starting to walk. He, of course, like most 1-year-olds learning to walk, fell on his bum at one point during our expedition.

We were watching from a safe distance to make sure he was all right. As he was in the process of getting back up and resuming his quest to do whatever, another parent swooped in yelling, "OH MY GOODNESS, POOR BABY, ARE YOU OKAY????!!!!!!????" She made a huge scene, gave me dirty looks for not immediately rushing in, and then said, quite loudly, as well, "WOW, YOU AREN'T EVEN CRYING." Yes, I know. Because when he started walking and falling down and then looking at my husband and me for reassurance, we said, "Whoopsie, you fell down!" and that was all. Now he falls down on purpose and yells, "I FELL DOWN," and laughs. Of course, if he's truly hurt, we are there for him immediately … but he's learned to brush off simple mishaps based on our response.

Infancy and toddlerhood also are the stages when the foundations for attachment are set. By attachment, I am referring to the psychological and biological phenomenon of forming a bond with a caregiver, not the attachment parenting trend (which is based on both very responsive parenting and physical contact). Infants are going to inherently engage in emotional relationships with their caregivers because it is more likely to lead to their survival. Hence, going back to Chapter 8 and evolution again. Both the initial foundational attachment and the continuing quality of the parent-child relationship are thought to be critical for all future relationships. Children who do not have the opportunity to form solid attachments, such as in the cases of emotional abuse or neglect, death of a caregiver, or lack of attention such as in the case of children in orphanages or foster care, can develop what are called attachment disorders. Attachment disorders, if left untreated, can significantly impact a child's future social and emotional relationships with others. The ultimate result ranges from impaired physical development to relationships problems, addiction, and anxiety/depression.

CHILDHOOD.

Unlike during infancy and toddlerhood, most of us have memories from our childhood and can remember various milestones, like losing our baby teeth and learning to read. Childhood is technically divided into two periods, early childhood, which spans years 2-6, and middle childhood, which is years 6-11. In early childhood, physical growth slows down and children begin to lose their baby fat. Children become more coordinated and can run, jump, and climb. By the time children reach middle childhood, they are continuing to grow taller, although there is less brain development during this phase than in early childhood.

Developmental psychologists have devised many clever (and occasionally humorous) experiments to test for various milestones in cognitive development during childhood. For example, theory of mind is something that develops during early childhood. Theory of mind refers to an organism's understanding that they have thoughts and emotions, and their thoughts and emotions may differ from someone else's. It also can be equated with the ability to think about thinking, or metacognition. Classic tests of theory of mind include the false belief task. In the false belief task, children are tested to see if they understand whether or not a false belief possessed by another can guide their behavior. In this task, children are presented with two boxes, one plain and the other marked, such as a Band-Aid container. They are asked to pick the box with the bandages, and they typically chose the box with the

Band-Aid markings. However, the researcher shows the child that that the box is empty and that the other box has the Band-Aids. Then the researcher presents the child with a puppet and asks the child where the puppet thinks the Band-Aids are. A child with theory of mind can articulate why the puppet might think the Band-Aids are in the marked box, thereby identifying the puppet's false belief.

Now, the really interesting thing about theory of mind is that it is not solely a human trait. In the mid-2010s, research came out demonstrating that corvids, a group of birds that includes crows and ravens, have theory of mind. Corvids cache or store their food. Researchers designed an experiment where captive ravens could look through a peephole to see where another raven was storing its food. They observed that if ravens knew the peephole to their enclosure was open, they behaved differently. This demonstrated that they both had some rudimentary understanding of what the peephole was for and knew what it meant if another bird was watching.[21] Research also indicates that primates, such as chimpanzees, as well as macaques, grey parrots, scrub jays, dogs, pigs, and goats, also have theory of mind.

Another interesting task to understanding human cognitive development is the marshmallow task. The marshmallow task is a test of executive function. Think of an executive like in an office: they are the one in charge. Executive function refers to the processes in the brain that control what we are doing. This includes things like being able to plan, pay attention, follow instructions, and multitask. It involves specific brain functions, such as working

---

21 Bugnyar, T., Reber, S.A., and Buckner, C. (2016) Ravens attribute visual access to unseen competitors. Nature Communications 7. doi:10.1038/ncomms10506

memory (manipulating information in a short period of time, like remembering which turns you need to make to get to the grocery store), mental flexibility (having the ability to shift your attention), and finally self-control (resisting impulsive actions).

How does the marshmallow task work? A young child is brought into a room with an investigator, and there is a marshmallow or other yummy treat sitting on a table. The investigator leaves the room and tells the child they can eat the treat, but if they wait 15 minutes until the investigator returns, they can have twice as many treats than originally was provided to them. Very young children have a hard time waiting, but as they grow and their executive function ability increases, they become more able to wait. Later experiments with the marshmallow task (this task was first developed in the 1960s) demonstrated that 4-year-olds who were capable of waiting also were more likely to grow up and lead successful careers.[22] Incidentally, videos of children participating in the marshmallow task can be quite funny to watch. You can look up these (or any of the other tasks mentioned in this chapter) on YouTube.

Swiss psychologist Jan Piaget had a background in biology, but is most well known for his work in cognitive development. Piaget thought there were four stages of cognitive development: sensorimotor (birth-2 years), preoperational (2-7 years), concrete operational (7-11 years), and formal operational (11+ years). One of the biggest contributions of Piaget's theories was in education, focusing on the child as an active learner engaged with their environment. However, later research indicated that these stages are

---

22 Mischel, Walter. (2015) The Marshmallow Test: Why Self Control is the Engine of Success. Back Bay Books.

not as clear-cut as was originally thought. Some skills develop later than others, skills can be improved with training, and the influence of culture on development was originally overlooked. There are some interesting tasks that Piaget developed, such as the conservation tasks. Conservation tasks are used to test if children can perform operations or logical tasks and come in many versions. In one example, the child is presented with two glasses of water, each in a similar sized glass (Figure 14A).

Figure 14. Demonstration of the conservation task.

The child is asked if there is the same or different amounts of water in each glass, and the child usually says there is the same amount of water. Then the experimenter takes one of the glasses and pours the water into a different sized glass and asks again if there is the same or different amounts of water in each glass (Figure 14B and 14C). A child who is preoperational will say there is a different amount of water, whereas the concrete operational child will recognize that the amount is the same.

Adolescence and Emerging Adulthood.

During my postdoctoral training, I remember sitting in on an undergraduate adolescent development course. My postdoctoral advisor was teaching the course, and I remember her introducing the period of adolescence by stating that it's a bit of a miracle any of us actually survive adolescence. It's an interesting period of time where many of the "safety" mechanisms that protected us while we were children are lost, but the inhibitory control and knowledge of adulthood hasn't yet developed. Think about it for a second ... can you come up with an example of something **incredibly stupid** you did while a teenager and you look back now and think about how lucky you are that you didn't get yourself killed or permanently mess up your life? It's all part of the normal developmental process. It fits in with an evolutionary perspective as well. As we discussed in earlier chapters, it is not ideal to reproduce with close family members. However, breaking away from a family unit that has provided safety and security for so many years also is risky. In order to create the best offspring possible, it makes sense that brain changes facilitate engagement in risky behavior (and also to be more peer-focused), because this is more likely to lead to successful reproductive outcomes.

Adolescence is the transition period between childhood and adulthood, typically during the second decade of life. It is also the greatest period of physical change that occurs during postnatal life. Hormonal changes trigger puberty marking the beginning of the physical transition to adulthood, which includes everything from reaching sexual maturity to growing pubic hair. Adolescents are becoming more logical and can reason more scientifically, although they still aren't as effective decision-makers as adults and can default into emotionality. In terms of social development, teens are seeking autonomy and peer relationships and begin to rely less on their parents, although a supportive and warm parental relationship is still important.

Emerging adulthood is a relatively new and controversial stage of development between adolescence and adulthood. Part of the controversy focuses on whether it is a real developmental stage since it has a tendency to vary between cultures and among socioeconomic statuses. In particular, emerging adults tend to be college students in their late teens and early twenties who still are not fully independent adults. During this period, identity development continues, there is an increased focus on preparing for a career, and romantic relationships can become more serious. One of the biggest risks of the emerging adulthood period is its "sink or swim" nature. Emerging adults will either make good decisions and successfully become independent adults, or they will remain stuck in limbo without an obvious direction, still dependent on their parents.

THE FOUNTAIN OF YOUTH.

Development never truly ceases. Even after you've hit adulthood, your body still undergoes developmental processes. However, at this point, it is instead called regeneration. The specific definition of regeneration given in Gilbert's Developmental Biology[23] (*the developmental biology textbook*) describes it as "the reactivation of development in postembryonic life to restore missing or damaged tissues." The molecular processes of development can repeat themselves later, except this time instead of making completely new tissues, existing tissues are repaired or otherwise rejuvenated.

For example, you have stem cells in your skin and digestive tract that are used to replace damaged or worn-out cells. The liver

---

23 Gilbert, Scott F. and Barresi, Michael J. (2017) Developmental Biology. Sunderland, MA: Sinauer Associates, Inc.

also has amazing regenerative capacity. Even if half of your liver is lost due to damage, it can regenerate. If a salamander loses a limb, it can regenerate the entire limb. Before I started working on my graduate degree, I was working as a laboratory technician. Our neighboring laboratory was studying zebrafish heart regeneration. They (and others) were able to demonstrate that damage to the zebrafish heart would induce the surrounding cells to de-differentiate (become more stem cell-like), divide to make more cells, and then re-differentiate or turn back into heart cells.[24] The reason regeneration could be likened to the fountain of youth is because theoretically, by knowing what to reactivate, you can continually replenish worn-out parts of your body, possibly extending the length and quality of your life.

Summary

Our development from a single cell to a fully functioning adult is a journey spanning both biological and psychological perspectives. From the biological perspective, we examined the mechanisms of how fertilization occurs and the genes and molecules work together at the right time and in the right place to properly form organs and tissue structures. We also discussed how the environment can influence development, both at the level of the genes and the overall brain architecture. Development from birth to adulthood varies among individuals. I like the popcorn analogy when considering children. You have a bunch of popcorn kernels in a pot over the stove, and then each pops at its own time. Parenting advice can get

---

24 For a review of this topic, see: Zuppo, D. A., & Tsang, M. (2019, September). Zebrafish heart regeneration: Factors that stimulate cardiomyocyte proliferation. In *Seminars in Cell & Developmental Biology*. Academic Press.

dubious for this exact reason. What works for one child may not work for another ... even if the children are siblings. There is no one-size-fits-all way to raising children. The biological and psychological processes involved in creating and raising tiny humans to adulthood are truly awe-inspiring.

In the next chapter, we are going to continue we are going to continue at the intersection between the intersection between psychology and biology, but in a slightly different way. We now will turn to a discussion of how psychology influences the ways people use evidence to reason about biological issues, particularly contentious ones like vaccination decisions and climate change.

CHAPTER 11

THE PSYCHOLOGY OF USING BIOLOGICAL EVIDENCE TO MAKE DECISIONS

I HAVE A needle phobia. Before you say, well, no one likes to get shots, I'll say it again: it's a phobia. When I need to get a blood draw or a routine vaccination, my body responds like I am in mortal danger. Since I unfortunately have a medical condition that requires routine blood draws, I have an entire system in place with the local hospital to get through venipuncture without vomiting, passing out, or having a full-blown panic attack. When I had a root canal a few years ago, the assistant commented on how it was the first time in her 15-year career she saw someone get a root canal without Novocain. I couldn't do it. Before you think I'm Hercules or something, the nerve was dead so I couldn't feel too much. Still, the discomfort was far preferable in my mind to getting a shot. It is neither rationale nor reasonable, and if I could do anything about it, I absolutely would. My hypothesis (which is consistent with scholarship on the development of needle phobias) is that it developed after traumatic experiences getting childhood vaccines and having too many of my baby teeth extracted (each extraction with its requisite Novocain shots).

Not surprisingly, my childhood trauma with getting vaccines turned into anti-vaccination leanings as a young adult. In other words, I was biased. *Anything* to get me out of getting a shot or a poke and I naturally latched on to it, no matter how disreputable the source. Not because the data was terribly compelling, but because I was so deeply afraid of it that any reason to not do it sounded great. I later learned to differentiate between my biased feelings on the subject and the medical evidence, and everyone in my family is up-to-date on their shots (including me). Here's the kicker: my story isn't unique.[25] Needle fears and phobias are fairly common and have major consequences on health and wellness in society. This includes outbreaks of vaccine-preventable diseases like the measles and increased health burdens because of a lack of routine medical testing or care. It could also impact society because people will avoid careers that require routine pokes. For example, when I was an intern at the Pittsburgh Zoo & PPG Aquarium, I had to have a tuberculosis skin test, which involves a poke! I was ready to walk away (I didn't, and I'm glad I *stuck* with it, no pun intended) just because of a simple procedure.

The other issue is having something stabbed through your skin is instinctually very wrong. Even without my traumatic experiences as a child getting poked, our deep-rooted biological instincts are going to tell us to not do things that break the skin. Therefore, getting a vaccine requires using the frontal cortex, the part of the brain involved in thinking, to realize that getting a shot will not in fact cause mortal danger, but vaccine-preventable illnesses can.

---

[25] See McMurtry, C. M., Riddell, R. P., Taddio, A., Racine, N., Asmundson, G. J., Noel, M., ... & Shah, V. (2015). Far from" just a poke": Common painful needle procedures and the development of needle fear. *The Clinical journal of pain.*

The other issue with weighing this evidence is that vaccines are victims of their own success. When I was a child growing up in the '80s and '90s, I didn't see my peers struck down by polio the way my parents or grandparents did. I've often wondered if we started calling polio by its other name, infantile paralysis, would people pay more attention to the devastating nature of the disease? We generally do not have firsthand experiences with these devastating diseases anymore, which combined with deep-seated fears of vaccines either from our childhood or our instincts, perhaps it is not surprising why it is easier for some to believe so-and-so's anti-vaccine blog than the peer-reviewed medical literature. I recently read an article[26] that discussed how in the absence of fear of disease, we instead fear the vaccine (or perceived threats of vaccination). Obviously, there are more factors here, such as barriers with science communication and overall science literacy (we'll come back to this in Chapter 14), but it definitely underscores the impact of bias when it comes to evaluating evidence.

Cognitive phenomena like bias or epistemological beliefs (beliefs about the nature of knowledge and where that knowledge comes from) influence how we learn, what we learn, and what we choose to do in our lives. Epistemological beliefs that science is a collection of facts generated by money-hungry scientists play into vaccine fear and other anti-science practices, too. As we touched on in Chapter 2, science knowledge is subject to change in light of new evidence, which can lend the impression that science knowledge is flimsy. Flimsy isn't the correct word, but tentative is—the power of science is that it can be revised in light of new evidence.

---

26 Drew, L. (2019). The case for mandatory vaccination. Nature 575, S58-S60.

This chapter is unique compared with the others. Rather than examine a particular biological phenomenon and its relation to our daily lives, this chapter considers the cognitive basis for how we reason about biology-relevant decisions, such as the decision to vaccinate. This chapter examines our beliefs about the nature of science knowledge and how various cognitive phenomena influence how we evaluate and make sense of science knowledge, with some commentary on best practices for evidence evaluation. As stated before, my goal throughout this book isn't to convince you to necessarily do one thing or another but to empower you to use evidence to make your own decisions—and in this chapter, my goal is to foster your ability to understand the psychology behind decision-making and give you the tools you need to evaluate and use evidence the next time you need to make a biologically relevant decision.

Epistemological Beliefs about Science.

What science is and how we reason about science knowledge are two interrelated aspects of cognition. We went in depth regarding what science is and isn't in Chapter 2. What we know about science is influenced by what we believe about science (and vice versa). What we know and believe is going to influence how we engage in scientific activities, such as generating an argument or using evidence to make a decision. This is important when considering education. Someone's epistemological beliefs about science are going to influence how well they learn in a classroom and how well they engage in scientific activities (or practices, as they are now called in the Next Generation Science Standards) like inquiry. As I briefly alluded to above, epistemological beliefs about science also influence how we make decisions in our daily lives.

What are epistemological beliefs? And what do they have to do with scientific decisions in our daily lives? Epistemology is a philosophical term referring to what knowledge is and how "we know we know." Within science, epistemological beliefs about science tend to fall into four categories: certainty, structure, justification, and source. Certainty deals with how confident we are that we know something and our recognition that science knowledge changes in light of new evidence. Structure of knowledge deals with how we connect knowledge in our brains. It refers to the theoretical construction of knowledge that we make and how we interconnect various individual pieces of knowledge. Justification is how we use evidence to support what we know. Source is where we get our knowledge from, such as a peer-reviewed journal or authority figure. We will go into each of these types of beliefs in more depth in this chapter.

Some of these epistemological beliefs about science are roughly correlated with nature of science principles. Nature of science principles are defined as the key principles that are used to distinguish science from other disciplines of inquiry such as religion or philosophy. For example, the tentative nature of science (how science information is subject to change in light of new evidence) and the lack of a universal scientific method are two such principles. In a weird quirk of the social aspects of science, people who come from science education backgrounds talk about nature of science, and people who come from psychology talk about epistemological beliefs about science. Some researchers (including me) say that these are two sides of the same coin, and as I mentioned above, what we know (nature of science principles) influences what we believe about science knowledge (epistemological beliefs about science).

Certainty of Science Knowledge.

Certainty refers to how confident we are in science knowledge. It is easy to sit in a biology (or any science) classroom and look at the gigantic textbooks and develop the belief that science is a collection of concrete facts. Otherwise, why would we have such gigantic books? This was illustrated in Chapter 4 when I discussed my ever expanding cell biology textbook. Science knowledge is inherently uncertain, because it is subject to revision in light of new evidence. We touched on one of the most famous examples of the revision of science knowledge in Chapter 2 when the geocentric model (the sun revolves around the earth) was revised in favor of the heliocentric model (the earth revolves around the sun). The reason people thought the sun revolved around the earth was also the result of firsthand bias, which we'll come back to later in this chapter. The firsthand bias was derived from the fact that if you go outside, you can see the sun move through the sky during the day. It makes sense that before the advent of telescopes and the discovery of the currently accepted evidence, people believed the sun revolved around the earth because we could *see it move* in the sky. The telescope gave people the ability to observe the heavens in a new way, and new evidence was gathered that suggested the earth revolved around the sun. The Catholic Church took particular exception to this, and the scientist who first put forth evidence of the opposite theory that the earth revolved around the sun, Galileo Galilei, was accused of heresy and died while under house arrest.

Practicing scientists (and perhaps ironically, many people I've interacted with who are religious) are comfortable with uncertainty. However, when the general public looks at "might's," "could's," and "maybes"—a class of terms that convey uncertainty called "hedges" by linguists—associated with science knowledge, it sends not the message that "yes, XYZ is well supported by evidence though it could still be changed in light of new evidence," but rather "scientists don't know what they are talking about." This is why when evaluating scientific evidence if you hear someone espousing absolute truth on the subject, you should immediately have a red flag about the trustworthiness of the source. Most scientists will tell you there is no absolute truth—we are constantly discovering new things. This is what makes science exciting. It isn't all of the stuff we know already (getting back to the textbook example in Chapter 4 and science education reforms, which we'll discuss in Chapter 14), but the uncertainty at the edges of human knowledge.

How does the concept of certainty feed into our daily decisions? In 1998, Andrew Wakefield published an article in the peer-reviewed scientific journal, *The Lancet,* proposing a link between the measles, mumps, and rubella (MMR) shot and autism. No one was able to reproduce his results, and later evidence indicated that he had falsified his data and the article was retracted from the journal. Retraction means that it is removed entirely from the scientific record and is no longer considered valid. However, if one believes that science does not change in light of new evidence, how do we make sense of a retracted journal article? Consider another hot button biological issue, climate change. There isn't perfect agreement on whether human-driven climate change is occurring or what the consequent impact is going to be. Since the financial impacts of climate change are particularly troublesome (there is that bias again!), it is easy to latch on to the hedges (those "might's" and "could's") used by scientists when discussing the overwhelming evidence of human-driven climate change as a way of discrediting this evidence—when in actuality, the presence of hedges is actually a good scientific practice.

Structure of Science Knowledge.

Structure of knowledge is sometimes called simplicity of knowledge. It refers to how much we perceive knowledge to be interrelated. For example, rising sea levels is one piece of information and on its own is fairly simple. However, when connecting rising sea levels to rising temperatures and why both phenomena are occurring, it becomes evident that these pieces of information are actually interrelated with one another. This is related directly to bias as well. Someone who believes scientific information to be straightforward also is more likely to consider only a single viewpoint (usually their own) and accumulate knowledge based on this viewpoint. It is more difficult to integrate information from other sources. Relating this to firsthand bias again, if I look eastward from my house, the earth really does look flat, because the Great Plains are ... flat. However, adding in other pieces of information, such as observations made by astronauts, it is clear that the earth is not flat.

Going back to our vaccination discussion, although as I mentioned before that I am not anti-vax, I do recognize there **is** a small minority of people that has very serious reactions to vaccines, and it isn't well understood why this happens. Per the Centers for Disease Control (CDC), it is estimated that about 1 in 1 million vaccines will result in some kind of a serious adverse reaction. Your chances of getting hit by lightening during the course of your lifetime is about 1 in 3,000 (being struck and killed is about 1 in 700,000). Your chances of dying in a car crash are about 1 in 100.

So, although serious vaccine side effects do occur and more research on who is affected and how and why this happens is needed, you are still far more likely to get killed in a car crash. One of my many issues with the anti-vaccine rhetoric is that this fact often is lost in the anti-science mania, and acknowledgement of this fact can result in others assuming you are also an anti-science person.

The overwhelming current evidence suggests that severe reactions to vaccines are very rare and it is far safer for communities to be vaccinated. It's the same as wearing a seatbelt in the car. Yes, in a very rare circumstance, such as your car going underwater or catching on fire, wearing your seatbelt could be a real liability. But the evidence and statistics suggest that you are far more likely to die in a car crash if you do not wear a seatbelt. The point is, if you're biased to one side versus the other, it is easier to incorporate information that fits your viewpoint and not make connections with any other evidence. Most controversies are just that—controversies—because there *are* valid arguments on either side, but people get so set on defending their viewpoints, they do not view the big picture of all of the knowledge that is present.

Structure of knowledge is related to another phenomenon called motivated reasoning. Motivated reasoning is how someone's underlying goals influence their judgments. Our underlying biases (and we all have them!) influence which forms of information we pay attention to and how we examine and evaluate that information. We are more likely to pay attention to something that supports what we think and ignore other information. Why do you think there is so much argument about Fox News versus CNN? I don't watch either but am well aware of how incensed people can get when talking about either news source, depending on their political affiliations. Studies also show that it is easier for people to remember information supporting their own viewpoints while also suppressing information that contradicts their desirable views. Furthermore, how people interact with information is influenced by their values, interests, identities, and the source of the information (as we'll talk about later on).

There is another cognitive phenomenon called epistemic motive as well. This refers to your personal goals for using knowledge. Are you trying to arrive at a single finite answer (are vaccines safe?) or to get at a well-rounded understanding of the issue (are vaccines safe **enough** and a better alternative to a preventable illness?). People trying to arrive at a single answer not surprisingly will wrap up seeking information early on when they feel they have found an answer (usually in line with their own biases), and not when they've fully examined all possible sides to a decision. This fits into bias as well, because people are less likely to adequately pay attention to information that runs contrary to what they believe.

Vested interests can play into evaluation of knowledge as well. If you are interested in a topic, or it has personal relevance to you (such as living in a coastal region that could be substantially impacted by rising sea levels), you are more likely to behave in a manner consistent with those beliefs. Recognizing my own biases here, is it really surprising that I've spent more time talking about vaccines in this chapter than other socio-scientific issues society faces today like climate change or genetically modified foods? I'm a parent to a young child living in a state with low vaccination rates—I breathed a huge sigh of relief when he was old enough to be fully vaccinated.

Relating back to climate change, overall acceptance of climate change is also roughly correlated with areas of the United States either most affected or most likely to be affected by it, including coastal regions and where I live near the Rocky Mountains.[27] Due to either firsthand bias (which I have from experiencing the various habitats between the plains and the alpine region of the mountains) or because of a vested interest (living in a coastal city), it is easier to accept the evidence for climate change.

JUSTIFICATION OF SCIENCE KNOWLEDGE.

Justification of knowledge refers to what aspects of science knowledge we accept and how evidence supporting that knowledge is generated through science. Science knowledge is generated through science inquiry and isn't the recipe-like inquiry the majority of us were exposed to in school. Due to various resource constraints (time, money, training, safety), most of what students see in schools (K-12 and higher education) is what is called "simple inquiry." Simple inquiry looks like a cookbook recipe. Get these materials. Make a hypothesis based on the pre-lab and your homework from last night. Put this on that. See the expected phenomena (Bubbles! Color changes!) and draw a picture. Make a conclusion that you saw what you expected to see. Count down the minutes until class is over. I remember when I was in college, I was in a four-hour cell biology lab section looking at cancer cells. I spent FOUR HOURS re-making observations that had been done hundreds (more likely thousands) of times. I was beyond frustrated at the utter waste of my time.

Simple inquiry gives people the wrong idea about science. It suggests that science is straightforward, generates facts, and is

---

27 This is a nice infographic if you'd like to view the data yourself: https://climate-communication.yale.edu/visualizations-data/ycom-us/

fairly lackluster. Authentic science inquiry, which is what real scientists do, is far more interesting. Someone told me when I first started graduate school that most of my dissertation would be based on results I generated during my last year or so, because the first few years are spent getting it all wrong. I thought they were being overly negative. But maybe they weren't ... in the four years it took me to earn my PhD, the results were mainly based on my last two years of work. Things go wrong in authentic science inquiry. Ideas turn into dead ends; roadblocks take months to get around; and the most random observations turn into fruitful avenues of inquiry. Working on the frontiers of new knowledge, there is no right or wrong answer because what we find is new. Making sense of that new information involves being creative (we'll talk about creativity and the arts in the next chapter) and making good arguments.

Since there isn't an answer key, we rely on the peer review process for external vetting of results. We briefly touched on peer review in Chapter 2 when discussing how science is done by groups of people. The idea is that others with expertise in a related field are the best judges of the evidence provided in support of new scientific insights. They can determine if it is valid or not based on the methods and interpretations presented by the authors. How does the peer review process work? When someone has a publishable body of work, they write it up as a paper and submit it to a journal for review. Journals vary in quality, and higher tier journals (like *Science*) have a more extensive peer review process. Usually the editor of the journal will examine the manuscript first and see if it is a good fit for the publication before sending it out to at least two, if not more, peer reviewers. Peer reviewers may or may not know the identity of the authors (the so-called double-blind peer review, which helps mitigate biases), and review the findings of the paper. This includes checking to see if appropriate methods are used, if the conclusions are supported by the results shown, if there is any evidence of academic misconduct (I have reviewed papers previously that were obviously plagiarized), and if the results presented are truly new to the field and not just new to the authors. If the two reviewers disagree, more reviewers often are brought in to independently review the work. The journal editor examines all of the reviews and then writes a meta-review and makes a decision about whether or not the manuscript is acceptable for publication. The manuscript typically isn't acceptable on the first pass—reviewers find something that needs to be addressed. Usually there is at least one, if not more, rounds of revision and re-review until consensus is reached that an article is, in fact, suitable for publication. Remember in Chapter 2, I had the screenshot showing how many times my paper was reviewed? I only went through one round of revisions before the reviewers felt that my article was acceptable for publication. You can also see when the article was submitted, returned for revision, and finally deemed acceptable. Quality research articles will often have this information.

I've heard several critiques of peer review as a biased process or that reviewers are "bought" by pharmaceutical companies to give certain reviews. This argument again gets into evidence evaluation and the transparency of the process. A journal that has double-blind peer review is less likely to have a biased review process. Journals that are transparent about their policies and how conflicts of interests are managed also are going to be more reputable. In either case, you can find all of this information online. Most reviewers (and authors) are required to disclose potential conflicts of interest, and this information is often published along with their article, so readers can make the final judgment call. A study on the effectiveness of XYZ funded and published by the company that makes XYZ is more than likely going to be biased. Funding information is always available in good journal articles, usually either at the very beginning or end. Also, I can tell you the exact amount of money I receive for reviewing papers: $0. Many scientists are pushing for this to be changed as reading articles carefully and writing thoughtful reviews takes hours—all of it uncompensated time. This particularly stings as some journals charge for publication, either as page fees or open access charges—and these charges can be in the thousands of dollars. So, the biggest issue with the peer review process is that scientists pay to have their articles published, the papers are reviewed for free, and the final product is often stuck behind a paywall (but not always, such as in the case of open access fees, which is when the author will pay to have their article freely available to the general public).

SOURCE OF SCIENCE KNOWLEDGE.

The source of knowledge refers to beliefs about where knowledge comes from. For example, you might be willing to believe everything that I say in this book simply because I have a PhD in molecular biology and lead an active research program in biology education. This suggests you have a belief in me simply because I am an external authority on the subject. It is easy to believe authority figures at face value without critically evaluating the evidence. Rest assured that this book has gone through a peer review and vetting process as part of publication (read the acknowledgements section book to see all of the various experts I consulted). It would be terribly easy for me to make things up and weave them in with the generally accepted information presented here to advance my own agenda.

There is also the additional consideration that people are far more likely to believe someone in their "in-group" as opposed to their "out-group," since this is hardwired into their psyche. In-groups are groups of people where we identify as members. For example, as a lifelong musician, I identify strongly with other groups of musicians and less with groups of athletes. I identify with other scientists and young parents, too. This also applies to gender identity, religion, family structure, and political affiliation. We are more likely to believe information if it comes from a source within our own in-group. I'm inherently going to trust someone who looks like me (either externally, internally, or both) more than someone who I perceive as different. This is why in educational settings there is a push for diversity among instructors to match their students—students will learn better from a teacher they perceive as part of their in-group.

Let's apply this to our vaccine discussion. I'm in a unique position where I identify with both biologists and other mothers as part of my in-group. Since I view scientists as part of my group, it is easier to trust scientific evidence around vaccines, and therefore, I decided to vaccinate myself and my child. For someone else who does not identify as a scientist, they are a member of an out-group, and, therefore, scientists can be seen as a less trustworthy source of information. A friend or family member's blog, coming from someone who is part of the in-group, seems a more trustworthy source, and therefore, it is easier to believe what they say instead, even if the evidence presented is biased and poor. We see this with firsthand experiences, too. Someone who has the firsthand experience of, for example, remembering the neighbor kid with polio, and shares that as a member of our in-group, is going to subconsciously be more trustworthy, even if one first hand observation is not necessarily the best evidence.

The reverse is true as well—another parent who has a child with a severe vaccine reaction, however rare the reaction was, will get a lot of attention from other parents simply because they are in the same in-group. The rarity of the reaction goes unnoticed, particularly with amplification over social media channels. This is the same phenomenon that was observed with the "milk carton kids." Images of missing children used to be on milk cartons, and this fed the belief that child abductions are very common, which then fueled enhanced stranger danger among parents. Now we have a society that talks about "helicopter" versus "free-range" parenting where you can have the police called on you for letting your child play alone at the park.

The other issue with firsthand experiences is that (a) we are all biased and (b) it is hard to make sense of information by itself. Biases actually can be so strong that they can cause someone to unconsciously change their behavior in order to fit their viewpoint. Since it can be difficult to make sense of information by itself, if doctors observe a strange phenomenon with a patient that doesn't seem to have any direct explanation, they will often publish what is called a case study. Case studies are detailed descriptions of a particularly unusual observation in a patient. If multiple people observe the same strange phenomenon, a meta-analysis may be performed that combines information from many sources to make some general statements about the phenomenon at hand. Evidence is gradually accumulated, which may result in changes in practice. On its own, a single instance or case study doesn't hold a lot of weight, but once connected with other pieces of information, it can tell a compelling story (this gets back to the structure of knowledge as well). Beware next time you see a horror story on social media about XYZ terrible thing that happened to one person. You don't know the context or the prevalence of the observation. Even then, if you find 10 people that something happened to, of the millions of people it could have happened to, the chances of it happening to you are probably still less than being struck by lightning.

Another consideration for source and evaluating claims that you read is to look at the source. At a basic level, some social media platforms like Facebook now allow users to learn more about the source of various claims to help them identify reputable sources. One thing to watch out for is "click-baity" or inflammatory headlines. Various news outlets want to get clicks to their website, and more interesting headlines tend to generate more clicks. One example used by my teaching mentor, Dr. Jacalyn Newman, was an article that equated getting up early in the morning to having heart attacks. The suggestion was we should all sleep in so we do not have heart attacks. However, who gets up early in the morning on a regular basis, teenagers or older adults? Typically older adults ... who, as a factor of their age are also more likely to have heart attacks. This is also a great example showing that correlation is not causation. Another example of the problems with thinking correlation and causation are the same is the relationship between rainy weather and carrying an umbrella. Carrying an umbrella does not make it rain outside even though rainy weather and carrying umbrellas are correlated with one another.

Interesting headlines can also grossly misstate or overrepresent the data. There are a few principles to keep in mind when evaluating claims you see in popular media. (1) Who wrote the article and where was it published? Is it in a reputable media source (for example, a well-known and respected publication versus My Neighbor's Blog)? Does the media source have known biases that could impact the reporting? Who is the author? Are they trying to promote an agenda? Are they making claims themselves or describing the claims of others? (2) If it is a popular media article about a scientific discovery, can you find the original sources? What are they? Good scientific reporting in the media will have links to the original, peer-reviewed research article. You may also see actual quotes from the study authors. (3) When reviewing the source article, is it published in a reputable peer-reviewed journal? Is the research conducted sound? Although this last question can be difficult to answer outside of the original study author's field, you can still look for conflicts of interest statements, the monitoring editor's information, the affiliation information of the authors, the number of people who participated in the study (and if it was a human subjects study at all. Remember the studies on essential oils I mentioned in Chapter 3? These were studies in cell lines, not humans, and so the generalizability to humans is questionable), and if the claims the authors make align with the claims made in a popular media article. You can often see other papers that cite the article in question as well and if there are any known controversies associated with the claims. For example, I've seen articles that are titled "[specific claim]: a response to [title of original article];" this way you can look at any controversies in the field for yourself and see other experts' opinions on the same topic.

Summary

I'm certain at some point throughout reading this chapter I've irritated you, at least sub-consciously, because I've mentioned at least one viewpoint that you found challenging for whatever reason. Here is another challenge for you: can you recognize what it was and why? This will tell you an awful lot about your own personal biases (again, we all have them! Myself included!). Making decisions about biological issues is a factor of what we know about how science works and how what we know then influences what we believe. Much of what we know and believe is going to be influenced by cognitive factors that we may not even be fully aware of. The process of evolution hardwired us to trust members of our in-groups, make quick judgments, and go with gut feelings. Modern science tells us to ignore this hardwiring, to attend to our biases, and to weigh evidence from a variety of sources, including those that run contrary to what we believe.

The core philosophy and goal of this book is to encourage thinking about biology in our daily lives—and that includes attention to how science really works. My hope is that this chapter gave you a better understanding of how our brains work when we are interpreting and using scientific information. The cognitive issues presented here factor into the final chapter of this book—the future of biology. Before we get there, we are going to examine two typically glossed over, but fascinating aspects of the practice of biology: the art of biology (Chapter 12) and the business of biology (Chapter 13).

CHAPTER 12

Everything I Needed to Know about Life, I Learned in Band: The Arts and Biology

THE IRONIC PART about the title of this chapter is that when I was younger and would daydream about writing a book someday, that was going to be the title: Everything I Needed to Know about Life, I Learned in Band. I'm a lifelong musician and feel quite old sometimes when I realize I started playing the flute more than 20 years ago. My interest in science has been around about as long (perhaps even longer). During that span, I have cultivated an interest in topics at the interface of science and music. I've noticed many times that the people who persist in music (in amateur groups at least) also tend to hold careers in the sciences. There are plenty of famous scientist musicians, including Albert Einstein who was a violinist. A dream research project I've had is to figure out why there is a music-science connection. Why do many people who gravitate toward science careers also tend to be musicians? If that is the case, why on earth are their either STEM (Science, Technology, Engineering and Math) **or** arts magnet schools for kids to attend? It bothers me that at face value, it looks like kids are forced to choose which one to excel at, when really the two are connected with one another. My experience suggests that integration of arts and biology and a genuine appreciation for each on its own is essential to the practice of science, an idea we'll return to at the end of this chapter.

Arts play a central role in biology. Not just because of creativity either, which we talked about in Chapter 2. It's also not arts for the sake of STEM, which is a critique of many schools' rationale for maintaining their arts programs. It's not for the sake of fostering better STEM outcomes among students and it's not art for art's sake (which, by the way is the motto of MGM—look above the lion next time you are watching a movie, where it says the Latin equivalent or "*ars gratia artis*"). Biological art is foundational to the field of biology. The artistic abilities of early scientists was essential for communicating observations on a variety of topics, including the structure of neurons, the changes that occur post-fertilization in an embryo, and the appearance of various animals in the field. Today, biological illustrators can be found creating art for patient education or surgical illustrations, creating visuals of molecular phenomena, and designing realistic prostheses. On the converse, our biology can explain our experience with the arts. For example, what is happening physiologically when we are emotionally moved by music?

This chapter will have two parts. First, we will start with an examination of the history of art in biology and the role of biological art today at a time when we can cheaply and easily generate photographs. Then, we will switch gears and close with the application of biological science to understanding what happens to us when we play or listen to music.

ART AS FOUNDATIONAL TO BIOLOGY.

The discovery that the brain is made up of individual cells separated from one another by neurons was foundational to the field of neuroscience. Neuroscience conjures up images of sophisticated and complicated technologies to understand the brain, and it may be easy to think that this key discovery also involved fancy technology. In fact, the key "technology" involved was Santiago Ramón y Cajal's ability to draw what he was seeing under a microscope. Cajal's freehand drawings of neurons can still be found in textbooks today. Cajal, the father of neuroscience and potentially one of the greatest biological illustrators, made his most significant contributions to science due to his ability to communicate his scientific observations through his illustrations. This is not an isolated occurrence. Prior to the development of cheap cameras, biologists relied on their ability to make fast, accurate drawings to record their observations and consequently to advance the field of biology.

It's not just neuroscience either. In developmental biology, which, as we talked about in Chapter 10, is the study of how organisms go from a single cell to a fully functioning organism, hand-drawn images using model organisms is integral to research. Model organisms are well-studied organisms that can easily be maintained in a laboratory setting and are useful for understanding biological processes. In the case of developmental biology, many of the biological "programs" that control the development of various organs or structures are evolutionary conserved. Since animals like zebrafish or African clawed frogs are externally fertilized, processes such as eye development can be observed in real time under a microscope. The zebrafish is particularly advantageous because the embryos are transparent, so you can easily watch internal organs like the heart develop. The development of model organisms is tracked in terms of stages (in the case of African clawed frog embryos) or the presence of key anatomical features (in the case of zebrafish).[28] Hand-drawn illustrations serve as the bedrock for teaching students how to identify various stages of organisms and for researchers to communicate their findings in the field. Such illustrations are still used today. In fact, to facilitate research on craniofacial development using the African clawed frog as a model organism, a group of researchers at Tufts University recently commissioned additional standardized illustrations, called the Zahn illustrations.[29]

If we have sophisticated and relatively cheap digital cameras, why bother with hand-drawn images? The answer to this question is actually quite complex. For one, there is the ability of art to reach people in a more personal way. Take Cajal's work for example. You do not need to be a neuroscientist to appreciate the beauty of these images. A colleague at the University of Alabama at Birmingham, Sarah Adkins, specializes in using agar art (Figure 15), which is creating art with multi-colored bacteria, as a method for increasing positive attitudes toward biology among undergraduate students.[30] She claims that the human element of creating an illustration is essential for communicating science concepts. Biological art, such as agar art that creates art from living organisms, namely bacteria, is a particularly important media for science

---

28 Zebrafish illustrations: Kimmel, C. B., Ballard, W. W., Kimmel, S. R., Ullmann, B., & Schilling, T. F. (1995). Stages of embryonic development of the zebrafish. *Developmental dynamics*, *203*(3), 253-310.

Original *Xenopus Laevis* illustrations: https://www.xenbase.org/anatomy/alldev.do

29 Zahn, N., Levin, M., & Adams, D. S. (2017). The Zahn drawings: new illustrations of Xenopus embryo and tadpole stages for studies of craniofacial development. *Development*, *144*(15), 2708-2713.

30 Adkins, S. J., Rock, R. K., & Morris, J. J. (2017). Interdisciplinary STEM education reform: dishing out art in a microbiology laboratory. *FEMS microbiology letters*, *365*(1), fnx245.

communication because of the use of life to communicate concepts about life.

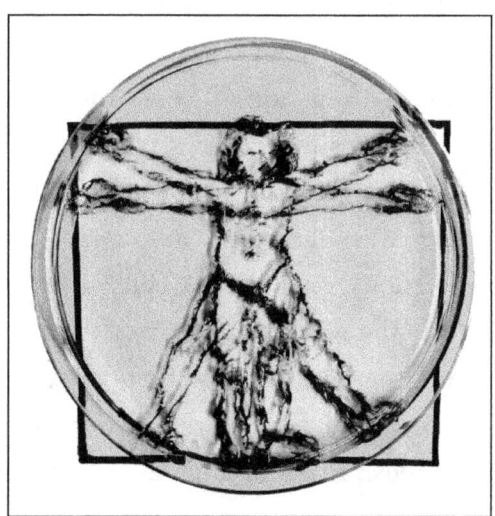

Figure 15. An agar art depiction of Leonardo DaVinci's Vitruvian Man, another famous example of illustration being used to represent a biological observation, this time anatomy. Photo courtesy of Sarah Adkins.

There also are plenty of biologically significant observations or phenomena that we cannot capture in photographic images; other times the images we take end up being substantially limited. For example, lacking a time machine, the only way to visualize lifelike reconstructions of extinct organisms is to generate illustrations or animations. Certain molecular phenomena are not directly visible, and either indirect information (for example, looking at cell extracts to see if a certain protein is present or not) or static black-and-white images, such as those generated through electron microscopy, are all we have for visualization. An artist can turn this information into an illustration or animation that is accessible for both scientists and the general public. Any decent textbook will contain illustrations. Sometimes these illustrations are simplified for the purposes of teaching a specific point. Other times illustrations remove distracting aspects. Our brains can only handle so much information (called cognitive load), and looking at an image of a dissected human body will include other information that is distracting (such as the presence of blood), whereas a beautiful anatomical drawing allows the viewer to focus in on one part, such as the structure of the bones and muscles in the arm. We see the same idea outside of biology as well with simplified road maps.

What about animations? Good animations also are routinely used in classrooms, especially to explain molecular phenomena. One of my favorite ways of teaching complex molecular interactions that we can't directly see with our eyes is to explain it in words, then with an illustration, and finally with a video animation. The repetition, particularly the repetition in various forms, really helps students understand complicated molecular phenomena. Good animations also can be useful for a scientist who wants to communicate new findings. Illustrations or animations are often included as part of peer reviewed journal articles. Animations also can be used to artistically describe how an extinct animal may have moved when alive.

In summary, illustration and animation in more recent times have been and still are important parts of the practice of biology. Art is important for communicating new discoveries to other scientists and also for teaching. The process of making new art also can lead to new scientific observations and avenues for research, which was the case with the Zahn drawings. Even more importantly, art is a way of making science accessible and relatable to all. Too often science and biology are portrayed as dull and abstract, and good art is an excellent way of bridging that gap and communicating key ideas to a larger audience in a more relatable fashion.

How else do we see biology and the arts interface? For the next half of the chapter, we will turn toward what biological and sciences can tell us about the experiences of listening to and making music.

PHYSIOLOGY OF MUSIC.

I've had a wide variety of interesting and moving musical experiences. I've performed before crowds in the tens of thousands while in a college marching band, been covered with goosebumps during critical moments of music, and done incredibly whacky things like play piccolo and hula hoop at the same time. The latter is immortalized forever on social media. I've also felt the undeniable connection with other musicians when making music together. It's not all in my head either—biologists have studied what goes on when people make music either alone or together and what happens in the brain when we listen to music. In fact, making music in particular has been linked to numerous cognitive benefits, including increased problem-solving ability, better executive function, and better memory. For musicians facing brain surgery, doctors will map brain functions related to listening and making music using functional Magnetic Resonance Imaging (fMRI) and use it as a guide during surgery. Even more, to ensure that none of the brain regions involved in making music were damaged, patients can be woken up from anesthesia mid-surgery and play their instrument as either a final check to make sure everything is working properly or to guide the surgeon during a risky part of surgery. Listening to and making music have differing effects on the brain. Now, we'll get into more detail of the biology of each.

Listening to Music.

When I was in graduate school, one of the local hospitals had a beautiful grand piano down in the lobby. You could sign up for a time to play. My husband and I would go—he would play piano, and I would play flute. Over the course of an hour, we would see doctors, nurses, patients, families, and children go by. Sometimes people would stop and listen. It often made people smile, especially with some of the more humorous pieces in our repertoire (including an arrangement done for our wedding of "So Happy Together" by the Turtles). We also played quite a bit with a community band at assisted living facilities throughout the area.

Why music though? Why is live music a good thing for people dealing with medical conditions or for those who are their caregivers? Why do we listen to music while we drive, work, or run? There is good evidence supporting listening to music while running—it is associated with improved running performance.[31] Additional research suggests that our musical memories are so durable, diseases that impact the memory, such as Alzheimer's, have less effect on the areas of the brain associated with musical memories. Said otherwise, these areas of the brain are less prone to neurodegeneration. Music is soothing to patients and caregivers. In fact, there is an entire discipline, music therapy, that provides evidence-based musical interventions. Musical therapists undergo special schooling in music therapy and engage in a therapeutic relationship with patients such as improving motor function in patients with Parkinson's disease or stroke victims, lessen the effects of dementia, reduce pain, and help premature infants sleep and gain weight.

What goes on in the body when we listen to music? When we listen to music we love, our brain releases the neurotransmitter dopamine. Neurotransmitters are chemicals released by neurons

---

[31] Bigliassi, M., León-Domínguez, U., Buzzachera, C. F., Barreto-Silva, V., & Altimari, L. R. (2015). How does music aid 5 km of running? *The Journal of Strength & Conditioning Research, 29*(2), 305-314.

to communicate with one another. Dopamine is important for regulating pleasure and motivation and is released during biologically important tasks like sex or eating. But why music? Music isn't key to our survival, so why does it activate reward pathways in the brain? It turns out dopamine is a key player in regulating pleasure associated with listening to music. Have you ever had chills while hearing a favorite song? This is the result of increased dopamine in your brain. Scientists have found that if you inhibit dopamine activity in the brain, there is less enjoyment from music and fewer chills when listening to pleasurable music. The opposite experiment of increasing dopamine in the brain increased chills and the overall enjoyment of the music.[32]

Making Music.
What about when we make music? The additional stimulation of the brain during the process of making music is even more complicated and interesting. I've heard making music described as a "full body" workout for the brain because many different parts of the brain, that of auditory, visual, and motor, are all active at once. Specific structural changes can occur in the brain that result from the specific type of instrument played by a person.[33] Evidence also suggests that encouraging young children to learn a musical instrument has a positive effect on their cognitive development and that structural changes within the brain can be

---

32 Ferreri, L., Mas-Herrero, E., Zatorre, R. J., Ripollés, P., Gomez-Andres, A., Alicart, H., ... & Riba, J. (2019). Dopamine modulates the reward experiences elicited by music. *Proceedings of the National Academy of Sciences*, 116(9), 3793-3798.

33 Vollmann, H., Ragert, P., Conde, V., Villringer, A., Classen, J., Witte, O. W., & Steele, C. J. (2014). Instrument specific use dependent plasticity shapes the anatomical properties of the corpus callosum: a comparison between musicians and non-musicians. *Frontiers in behavioral neuroscience*, 8, 245.

observed after 15 months of musical training.³⁴ In particular, we see that children who play the keyboard once a week, compared with children who go to a general music class with singing, dancing, and basic percussion instruments, had better motor and auditory skills.

What about the response to making music in a group? The connection musicians perceive between their personal experience making music and the experiences of those around them is documentable. Scientists in Sweden examined the relationship between singing and heart rate synchronization. Deep breathing slows down your heart rate. This is why if you go to the emergency room at a hospital in the throes of a panic attack, the first line of defense is intentional deep breathing. For those of you reading this who are Catholic, one cycle of the rosary takes 10 seconds to complete and causes an individual to take a breath approximately once every 10 seconds, stabilizing cardiac rhythms. Anyway, back to the study. The authors found that people who sing together are not only breathing in synchrony, but their heart rates accelerate and decelerate simultaneously as they sing.³⁵

Although it was not examined by the authors of this study, I would predict that a similar phenomenon would be observed among instrumentalists as well. Many conductors will begin a piece of music with a group in breath prior to the down beat to "link" all of the musicians together. When I play with a piano

---

34 Hyde, K. L., Lerch, J., Norton, A., Forgeard, M., Winner, E., Evans, A. C., & Schlaug, G. (2009). Musical training shapes structural brain development. *Journal of Neuroscience, 29*(10), 3019-3025.

35 Vickhoff, B., Malmgren, H., Åström, R., Nyberg, G., Engvall, M., Snygg, J., ... & Jörnsten, R. (2013). Music structure determines heart rate variability of singers. *Frontiers in psychology, 4,* 334.

accompanist, rather than counting off, we make eye contact and breathe together to ensure synchrony when playing. Interestingly, studies also demonstrate that people who sing together bond more quickly with a social group than those who do not, suggesting that encouraging people to sing together is an effective ice breaker activity.[36] This makes some evolutionary sense, as cohesive groups of early humans who felt connected with one another were also more likely to work together and consequently survive.

THE STEAM MOVEMENT: INCORPORATING ARTS IN SCIENCE EDUCATION.

I mentioned at the beginning of this chapter that some view science and the arts as polar opposites, or in terms of one versus the other. In recent years, educational reforms focus on the STEM subjects. The Next Generation Science Standards (NGSS) were developed to improve science education in the United States. This makes sense given that the current rate of college graduates in STEM fields isn't predicted to be enough to meet the needs of tomorrow's workforce and how important STEM jobs are for the economy. Buying kids STEM toys is very popular right now, too. It seems that if it's a STEM toy, it must also, therefore, be better than a non-STEM toy. Many of these toys involve more open-ended creative play elements and even could be considered more "old-fashioned" since kids have been playing with blocks for generations.

Really, what makes it a STEM or non-STEM toy? Sometimes it's because the toy is inherently focused on some kind of science-y

---

36 Pearce, E., Launay, J., & Dunbar, R. I. (2015). The ice-breaker effect: Singing mediates fast social bonding. *Royal Society Open Science*, *2*(10), 150221.

thing, like kitchen experiments or finding critters in the backyard and identifying them. There are also lots of games on the market now that teach kids how to code. In addition, there are inherently creative toys that are variations on a theme of open play that has been around for generations. For example, you can still purchase plain wooden blocks for kids to play with. Or you can buy the expensive version of the wooden block, Magna-Tiles (Figure 16).

Figure 16. My son building with Magna-Tiles.

Magna-Tiles are a building toy with magnets so you can build even more elaborate structures. They aren't kidding about the 3-99 age rating on the box either. I enjoy playing with them as much as my toddler does. The point of these STEM-branded toys is they help foster creativity through open-ended play, and creativity is critical for success in the STEM fields. As we discussed in Chapter 2, creativity is required to design experiments and interpret results.

Reiterating a point from earlier in this chapter, the arts, specifically the creation of illustrations and animations, is an essential part of science practice as well. Biological art is important for identifying new areas of scientific research, communicating findings to other scientists, and teaching students about the subject. It isn't terribly surprising that so many people are now talking about STEAM instead of STEM. Now we have the A for arts.

What does that A for arts really mean when we think about learning biology and encountering biology in our daily lives? Is the A for arts because biologists need to use the arts to be successful in their work, for example being able to write up their findings coherently? Is the A for arts so biologists can have enough rudimentary drawing ability to be able to quickly sketch out a surreptitious observation? Or is the A just an excuse to go out and buy arts and crafts supplies and justify purchasing a 3D printer, a mainstay technology in many STEAM maker spaces?

It is an interesting time to be a biology educator (a topic I'll return to in the final chapter of this book) in light of all of the recent science education reforms and trends. Much like I argued at the beginning of this chapter, true integration of biology and the arts comes from a fundamental appreciation for each field alone, plus the appreciation for what we can do when we combine the two together. This kind of interdisciplinary perspective is something I'll return to in depth in the final chapter. I don't agree with the arts for the sciences' sake or that somehow the sciences are more important than the arts. Or that the arts should only be taught because it helps with something else. For example, there is overwhelming evidence that playing music fosters cognitive development and helps kids grow up to be more successful adults. It's great to involve kids in music programs for these reasons *and* because there is certain joy associated with creating and sharing music.

Earlier in this chapter, I mentioned research showing the physiological changes that occur in individuals and across groups and how that tells us the arts are more than just a means to an end to promote science education. Art for the sake of art makes us better scientists and people. When we have an appreciation of art for art's sake, we can incorporate that perspective into new scientific discoveries and develop and enhance science communication strategies. This book is a direct result of that. Writing about science in an accessible manner is an art and something I very much enjoy doing (and, one could say, something that I'm also good at). Yes, as a scientist I also need to be able to write well in order to get my research published and convince others why they should fund my research (arts to an end), but as I'm sure you've noticed by now, this book is using writing in a completely different manner. Rather than using arts to an end, it is a creative manifestation of my love for biology and sharing that love with others. We see passion for the arts and sciences in agar art (Figure 15) and in continued efforts by biologists to maintain hand-drawn illustrations in the process of biological research and science communication. Yes, arts education will help make better scientists, but for its own sake and not just as a means to an end.

Summary

In this chapter, we viewed the intersection of the arts and the sciences through several lenses. We opened with a discussion of how biological illustrations have been used in various biological disciplines and how one scientist's illustrations revolutionized the field of neuroscience. The ability to use the graphic arts to communicate biology concepts continues to be important for both the process of science and the translation of information to others. Biology as a field also can tell us about how we, as humans, experience the arts on a physiological level, both as individuals listening to and creating music together. We closed with a brief discussion of the relationship between the arts and science and the potential pitfalls of emphasizing one field at the expense of the other. Now we will turn our attention to a different interdisciplinary application of biology to another field. In the next chapter, we'll look at the relationship of business to biology.

CHAPTER 13

FINDING NEW TREATMENTS: THE BUSINESS OF BIOLOGY

PRIOR TO WRITING this book, I was struggling with a significant and complicated health problem. After trying several treatment options, my provider recommended I try a newer medication that was still under patent. Getting that prescription filled was a nightmare. My provider had to send a letter to the insurance company to get a preauthorization, including details of the other medications I had tried and that did not work. Insurance companies want to take the cheapest route possible, so documentation of already trying the cheaper options was necessary. The pharmacy didn't have the medicine on hand because it was so expensive. In fact, they wouldn't order it until the insurance company first approved it. Once all this happened, I could actually go pick up the prescription (about a month after my doctor recommending it). Even with my insurance, I paid $2 per pill for a daily medication that without insurance cost $10 per pill. The worst part was all that work was for nothing. I had such severe side effects that I had to discontinue it after a few weeks, and the rest of the very expensive prescription went to the police department on a Drug Take Back Day.

Why the expense and hassle to get my meds? The medication was still under patent. Said otherwise, the formulation was not released yet, so no generic options were available. Developing new medications (or any other medical device or treatment) is very expensive and time intensive. It is estimated that it can cost in excess of $1 billion to develop a *single* treatment and/or medication. Each of the four phases of a clinical trial (we'll come back to what happens in each phase later in this chapter) can costs millions of dollars, and each subsequent phase typically costs more than the last. The biggest drivers of the huge cost of conducting clinical trials comes predominantly from the costs of procedures and medicines, staff to oversee the trial, and the monitoring of data collection at the hospital or other setting where the clinical trial is occurring. Since the majority of compounds tested won't actually make it to the final stage of approval by the Federal Drug Administration (FDA) (and consequently to market for patients to use), the exorbitant resulting prices of the ones that do are necessary to recoup the investment and to fund the research and development of other new products.

How does this process work? What does grant money have to do with it and why do certain diseases receive more research attention than others? Here, we will look at the business of biological research and how money flows drive research directions and treatment development.

How Does Biological Research Work?

Research in biology generally falls into two categories: basic and translational (also known as preclinical) research. Basic research is designed to answer fundamental questions about how life works.

This can be done in cell lines (cells cultured in a dish) or with model organisms, which we talked about in Chapter 11. An example of a basic research project may be to understand how a certain population of cells responds when treated with a certain chemical. In my own work in graduate school, I looked at how gene expression in neural stem cells from a male versus a female responded differently to treatment with a clinically relevant compound, dexamethasone. When we consider other biological research, such as ecology or evolutionary research, the majority falls under the basic research category as well.

Basic biomedical research (and educational studies as well, where the final application is in the classroom rather than the clinic) leads to translational/preclinical research and potentially clinical research, which is research performed in a medical setting, such as a hospital. Translational research takes insights from basic research and applies them to larger, more complex models in preparation for moving to clinical research. For example, basic research may have revealed that a certain cellular process is important for promoting uncontrolled cell division and cancer. Translational research would test a compound that inhibits this process, first in cells and later in animal models, to determine possible dosing. If the compound is promising, it may then make it to the clinical research stage and be tested in humans (more on clinical trials later in this chapter). As mentioned in the introduction, from start to finish, this process is extremely expensive and spans many years.

Who pays for this very expensive biomedical research and who does the work? Typically, biomedical research occurs in either university or industry laboratories. At universities, work is typically performed by undergraduate or graduate students and postdoctoral fellows. Postdoctoral fellows, or "postdocs," are individuals who have earned a doctoral degree, but are seeking additional training and experience prior to getting a permanent position in either industry or academia. Although I could write an entire chapter on the postdoc problem in academia, particularly in the biomedical sciences (which some have gone so far as to call a "postdocalypse," or as I mention with my students and mentees, the postdoc "black hole"), suffice it to say much of biomedical research is performed by these underpaid and underappreciated scholars all competing for very few permanent positions. Some research is performed by permanent researchers, such as research associates or professional research assistants. They mainly do research, either their own or someone else's depending on the funding sources. They may or may not teach as well. In industry positions, the majority of researchers are in stable, permanent positions that pay better, and there is less research performed by trainees.

In private settings, research is typically funded through product sales. In university settings, the research is funded through private or federal grants. Private grants come from foundations, either started by wealthy individuals (such as the Bill and Melinda Gates Foundation) or focused on a specific condition or disease and typically funded through donations (like the American Cancer Society). Federal grants come from taxpayer dollars. In terms of biomedical research, the National Institutes of Health (NIH) is the major funder. Basic research and research dollars for the other sciences and educational research comes from the National Science Foundation (NSF). Additional governmental entities, such as the Departments of Defense, Education or Agriculture, also will fund various research projects.

Each agency has a variety of grants available in various programs. NIH is made up of 27 different sub-institutes or centers. Each institute funds research in a specific area. For example, the National Cancer Institute (NCI) focuses on research on cancer, and the National Institute of Mental Health focuses on research on mental health. NSF is eclectic and includes various directorates, which are different centers or departments. NSF funds research in many science domains (including educational research) as well as programs specifically for improving undergraduate education, holding workshops, and developing and studying educational technologies. Within each agency, there are a variety of grants available depending on the stage of a scientist's career. For example, NIH has training grants that are specifically designed to support either a graduate student or postdoc during their studies. These often cover full tuition for students, in addition to providing a modest salary and health insurance benefits. The bulk of awards are for university faculty, senior researchers (such as research associates), or other qualified professionals outside of academia (for example, a researcher at a museum).

Funding Priorities: What Gets Funded and Why?

If someone has an idea for a research project, they begin looking for grant funds to support their work. One of the odd things about the scientific enterprise is that research tends to fall under the funding priorities of each specific agency. While this makes sense, it can create a somewhat biased representation of research dollars, and certain projects, such as interdisciplinary work that doesn't fit into a specific box or directive, are less likely to get funding. In the case of a disease like breast cancer, since it impacts lots of people, it makes sense that there are so many agencies and foundations that specifically fund breast or other cancer research. As a result, there are decent treatments available and new advances being developed all of the time.

What about rarer diseases? Rare diseases are those that affect few people. In the United States, the NIH's definition of a rare disease is one that affects less than 200,000 people a year. However, estimates of individual rare diseases are around 6,500 to 7,000, and the total number of people in the United States with a rare disease is estimated to be between 25 to 30 million. The problem with rare diseases is because they aren't common, there is not a ton of advocacy or research dollars for finding treatments … and consequently there are very few FDA-approved treatments. From a pragmatic standpoint, a million dollar grant that will help fund treatments for millions of people seems better than spending a million dollars to develop treatments that will only impact thousands of people. However, many of the rare diseases identified impact children and are life-threatening. There are several advocacy groups working to change these statistics, and in 2019, NIH announced several new grant awards forming a Rare Diseases Clinical Research Network.

Sometimes rare diseases will receive quite a bit of attention (and consequently research support). For example, *Life According to Sam* was a documentary produced by HBO in 2013. The documentary (which also received several accolades) focused on Sam, a young man with progeria, a rare genetic disease that affects only a few hundred people worldwide and causes accelerated aging. Sam's family founded the Progeria Research Foundation, and the money they were able to raise funded research that uncovered the gene that causes the disease and identified a potential treatment option.

Sam died a year after the documentary was released at the age of 17 due to complications from progeria.

Another grassroots example was the grant received by the Hannah's Hope Fund by PepsiCo. PepsiCo funded a grants program in the early 2010s. The company's goal was to fund new ideas that would have some kind of measurable positive impact at the community, state, or national level. Ideas could be submitted online, and people could vote on which ideas they liked best. As part of a marketing strategy, people who bought Pepsi products could get extra votes by entering in a code found on the cap of their bottle of Pepsi. One of the company's awards was to the Hannah's Hope Foundation. Like the Progeria Research Foundation, Hannah's Hope Fund was started by the parents of a child with a rare genetic disease. The foundation raises money to support research on Giant Axonal Neuropathy (GAN), a rare disease that causes progressive degeneration of neurons that starts in early childhood. Affected children typically slowly lose muscle control and die in their teens or twenties. There is currently a GAN clinical trial going on that started in 2015, and Hannah Sames (the same person the charity is named for) received treatment as part of the trial. According to the foundation's website, all of the preclinical work leading up to the trial was funded through the Hannah's Hope Foundation.

Applying Basic Biological Research: Human Subjects Research and Clinical Trials.

For the last seven years, I've been doing human subjects research. Really, I thought my graduate work in neural stem cells was tough because the cells have a tendency to be heterogeneous, meaning quite different from each other. But it was nothing compared to the process of doing research on humans. There is so much variability among the human population! Biomedical scientists also do not have the best reputations historically for ethically conducting human subjects research, so the procedures involved with conducting any human subjects research, and particularly biomedical research, are now very extensive to protect research subjects.

Anyone who wishes to conduct research with humans is required to undergo training in human subjects research. This training covers the history of human subjects research and what federal protections are in place. It includes reviewing the Belmont Report, which summarizes the ethical principles and guidelines for conducting human subjects research in the United States. Part of the motivation for the Belmont Report (and the establishment of the Office of Human Research Protections) was in response to the highly unethical Tuskegee Syphilis Study. In the Tuskegee Syphilis Study, 600 poor black men were enrolled and infected with syphilis (after being told they were receiving free health care) for the purpose of monitoring the progression of the disease. The participants were not told they were being infected with syphilis and were not offered antibiotics to treat the disease. A whistleblower in the early 1970s led to an end of the study. As a result, there is now an informed consent process that explicitly lays out all test procedures and rationale for doing so and requires that all research involving human subjects be reviewed by an independent Institutional Review Board (IRB).

So now if I wish to conduct a research study involving humans (even if it's an educational study), I need to prepare what is called an IRB protocol that describes my research rationale, objectives, and procedures in exacting detail. I also need to submit with the IRB my informed consent/assent documents and advertising materials. Research involving minors involves an assent process (the child has to agree to participate) and a consent process (the parents have to agree for the child to participate). Both processes involve different IRB approval forms. Someone in the IRB office will review the materials and request changes until the office is satisfied that my research project is being conducted ethically. Then I'll receive a letter granting me approval to conduct my research, usually for the span of one year before I will need to file an extension.

Another consideration for doing human subjects research is if any of the participants belong to a vulnerable group. Sometimes it is not ethical to do research on vulnerable populations. This includes people who are not able to fully consent (such as minors or individuals with dementia), people who are incarcerated and could feel coerced to participate, or pregnant women, since anything the mother does can impact the fetus and the fetus is unable to give assent to participate. Research on these specific populations usually involves inclusion of additional protections as part of the IRB protocol for the sensitive groups. An artifact of the challenge with researcher on vulnerable populations is one I mentioned in Chapter 10. Very few medications have been tested for use in pregnant women due to concerns for the unborn child.

If after extensive preclinical testing in cell lines and model organisms indicate that a certain treatment option, be it a medication, device, or vaccine, is promising, the intervention will then be tested in humans. The purpose of clinical trials is to determine if something is safe for use in humans and if it works as intended. There are four parts or phases to a clinical trial, and we will go through what happens in each phase in order. Each phase has slightly different aims and enrolls progressively more people. The first phase is screening for safety and typically includes less than 100 people. Phase two determines how well a treatment modality works. This can involve double-blind randomized control trials (RCTs), depending on the type of treatment tested. Double-blind means neither the patients nor the doctor knows if the patients are getting the experimental drug (experimental group) or placebo (control group). This is done to eliminate possible bias (which, as we talked about in Chapter 2 and Chapter 12, is not always conscious, but can influence behaviors, decisions, and interpretations). Although double-blind is preferred, sometimes it is not possible to do, for example if the comparison is between receiving speech therapy or not. Trials that are testing new drugs are typically double-blind. For example, let's say a group of investigators want to test if a new anti-depressant medication improves depression and anxiety symptoms. Research subjects involved would be randomly seeded into either the control (placebo) or the experimental (treatment) group.

The placebo effect is an interesting example of bias. The placebo effect occurs when someone thinks they are getting treatment for something and their symptoms improve. A person's mind can trick them into believing that the placebo (you could think of this as a fake treatment) is actually having an effect. Depending on the study, the placebo could be plain water, salt water, or a sugar pill. If someone expects a treatment to work, it does. The really interesting thing is if someone expects side effects, they get side effects (called a "nocebo effect"). This is why it is so important to compare treatment to placebos and that the individuals do not know which group they are in.

Randomized means that a participant is assigned to the control or experimental group at random. For example, patients may be assigned a random number by a computer, and then these numbers could be drawn out of a hat to assign the participants to groups. Randomization is typically done by someone (or a team) who is external to the experiment and has no contact with the patients. The team typically oversees the clinical trial, knows which patients belong to which group, and collects and organizes the data generated by the medical team. Their oversight is important for determining if the experimental treatment raises any safety concerns that may require the trial to be adjusted or halted early. Occasionally, there are treatments that work so well that it is considered unethical for the trial to continue. For example, if an experimental cancer treatment is incredibly effective at treating cancer with minimal side effect profiles, a double-blind RCT may be halted because it would be unethical to continue giving the other cancer patients placebos and not the highly efficacious treatment. There are also times where unintended side effects and intended use change the consequent marketing of a treatment. For example, sildenafil (more commonly known as Viagra) was initially developed as a treatment for high blood pressure. During clinical trials, it was discovered that it wasn't terribly helpful for treating blood pressure ... but quite good at causing penile erections. As a treatment for erectile dysfunction, it became hugely popular. So, the intended treatment was a flop, but the side effect ended up being the big seller.

RCTs are considered the "gold standard" of research with humans, in both medical and educational settings, and typically involve hundreds of participants. This is because they are able to minimize bias as much as possible while generating important results. This is why looking for FDA approval on something before trying it is important. Many natural or homeopathic remedies often have a disclaimer stating that the effectiveness of the claim made has not been rigorously tested and independently evaluated and ultimately approved by the FDA. This, of course, is biased as well since a natural compound like linalool (which as we talked about in Chapter 3 is the main chemical component of lavender essential oil) is not patentable and is never going to generate enough income to justify the expense and time commitment involved in an RCT. This also gets back to the point that with no evidence, we can't say it is or is not efficacious either, and it comes down to making the best decision you can with the evidence at hand. The time and expense necessary to perform an RCT is a major downside and certainly plays into why things that are tested clinically are also the most likely to lead to a return on investment. RCTs can take years to complete, especially if the condition studied is rare. Since they take so long to complete and involve large teams of patients and their caregivers, physicians, nurses, biomedical researchers, and statisticians, they are also not surprisingly very expensive, typically in the millions of dollars per RCT.

The third phase of clinical trials enrolls thousands of people. The reason the number of people enrolled continues to go up is to generate statistical power. Without getting into an overly detailed statistical description, the more people (or N) a study has, the more "powerful" it is, because there is less of a chance that what is observed is due to random chance. Going back to Chapter 11 and the notion of evaluating evidence, this is why when reviewing a study, it is important to look at the number of people studied. The fewer people involved, the more likely it is that an observed effect is due to random chance. By enrolling thousands of people, if the data continues to support that the benefits of the treatment outweigh the costs, this strengthens the argument of it being a good treatment. It shows that the treatment is effective at doing what it is intended to do.

In phase four, after the treatment receives FDA approval and goes to market, additional studies continue monitoring the risks and benefits to ensure that with an ever increasing sample size, the treatment is still acting as intended, with benefits continuing to outweigh the costs.

Harkening back to Chapter 2, the continuation of collecting data after FDA approval demonstrates that scientific inquiry is never over. Everything is subject to change in light of new evidence. Occasionally FDA-approved medications are recalled after approval because ongoing safety reviews indicate that there may be undesirable safety characteristics. This is why certain medications may be a better option clinically because of how long they've been around. A medicine that has been used for decades also has decades of safety and efficacy data as well. The converse is that it may not work as well as the newer treatments or medicines that have been designed to replace it. For example, conventional Selective Serotonin Reuptake Inhibitors (SSRIs) are the frontline antidepressant and anti-anxiety medications, and the best studied compounds work through a single receptor called SERT, or the serotonin transporter. Some individuals have a mutation in this receptor that prevents these medicines from working and so are forced to turn to other less desirable medicines that may have more severe side effects or that are newer and have less safety data. Even in patients who do not have the mutation, they can also have undesirable side effects from SSRIs. So, there is research on developing medicines to treat anxiety or depression that work through different pathways in the brain so more people can use them with fewer side effects. The bottom line is that data from RCTs is about as trustworthy as we can generate and it's tough to get FDA approval, but even with all of those safeguards, things are still missed. The important part is that the process of science allows for this, since we continue to collect data and revise our conclusions in light of new evidence.

Summary

In this chapter, we explored the business side of biology, or how money flows influence what gets studied and why. We discussed the nature of biological and clinical research, as well as who does the work and how this work is funded. We talked about federal, private, and grassroots efforts to fund biomedical research and how clinical trials work. New treatments are very expensive because the cost of developing them and then consequently performing safety and efficacy testing is laborious and expensive. However, the data generated (and the data that continues to be generated) by these processes gives us some confidence as consumers that the FDA-approved medicines and treatments we seek actually do what they are intended to with minimal, dangerous side effects. We also touched on how money directs research, either by foundations that seek donations to support research on a particular disease or traditional remedies being less likely to be studied as thoroughly because it is not possible to recoup the costs.

The business of biology impacts all of us and our consumer decisions, and as stated in the introduction, part of the goal of this book is to empower you to make informed biological decisions in your daily life. How do we empower others to make informed biological decisions as well? As we close the penultimate chapter of this book, we now look to the future and explore a few questions about the changing face of biology education and biology education research.

CHAPTER 14

NEXT STEPS AND BIG IDEAS IN BIOLOGY EDUCATION

AS I MENTIONED in chapter 1, one of the biggest motivators in writing this book is to connect with those who have had a poor experience in biology or other science classes and are willing to give engaging with this content a second chance. My goal with writing *Biology Everywhere* is to foster interest in these topics and instill confidence in you, the reader, to engage with the biology in your daily life. I've run into far too many people who have had terrible experiences in science or biology classes and consequently lack the confidence, interest, or both in engaging with science. It seems all too common that what we learn in school is disconnected from our daily lives. We are at a critically important juncture in the applicability of biological concepts to how we live. As a society, we are facing several biological issues that impact us all, while rampant science illiteracy often prevails. It is more necessary than ever for all of us members of society to be able to grapple with and make informed decisions about genetic technologies, medical and health decisions, climate change, and conservation.

Biology education (really, all science education) is in the midst of major reforms at both the K-12 and higher education levels. In K-12 education, the Next Generation Science Standards were released in 2013 to create a common standard curriculum in the United States and thereby (hopefully) to promote science literacy. In higher education (and also now bleeding into K-12 classrooms), reform efforts are changing science education from the traditional "sage on the stage" model, where a teacher talks and the student listens, to fostering methods of teaching and learning that are both evidence-based and student-centered, empowering students to actively construct their own knowledge, also known as active learning. Rather than students passively listening to the teacher, classrooms are more focused on allowing students to actively build their knowledge. For example, students may work in groups to solve some ill-structured problem or a problem without a definite answer, also called problem-based learning. Or they might study a case that has some underlying mystery and associated learning objectives (case-based learning) or complete a project (project-based learning). These can also be smaller scale efforts, such as think-pair-share activities, when the teacher poses an open-ended question and students first reflect on the question, pair with a partner to discuss their ideas, and then come back and share with the group. Since the publication of a large meta-analysis in 2014[37] demonstrating that active learning strategies improve student learning in science classes, there have been several calls for research to better understand who these strategies work for and why. This necessitates collaboration between biologists interested in education research and learning and scientists who study learning from cognitive, developmental, or educational perspectives.

---

37 Freeman, S., Eddy, S. L., McDonough, M., Smith, M. K., Okoroafor, N., Jordt, H., & Wenderoth, M. P. (2014). Active learning increases student performance in science, engineering, and mathematics. *Proceedings of the National Academy of Sciences, 111*(23), 8410-8415.

After reading this book and noting the presence of several chapters that cross disciplinary boundaries, it may not be surprising to you that I am one of the researchers sitting at this juncture between biology and psychology. I'm trying to better understand how people learn and understand biology. The rationale is that I can combine my deep content knowledge about biology, my understanding of the enculturation process of becoming a biologist, and my experience teaching biology with students ranging in age from preschool to senior citizens, with my understanding of how humans learn and develop from psychology. Although it makes good sense to approach how people learn about biology from this perspective, and there are several people who take this research approach, it is overall not very common.

There are many reasons for this. For example, each discipline traditionally approaches research from a different perspective. Even within disciplines, there is variability. In psychological research, people will argue about the validity of qualitative, meaning broad detailed observations of a phenomenon geared toward developing new theories, typically with few research participants, versus quantitative methodologies, which typically involve large numbers of research participants with some kind of numeric output, such as performance on a survey or grades in a course. There is dispute over which methodology is "better" ... when really, appropriateness of methodologies is better determined by the research question asked (which we touched on in Chapter 2). Qualitative methods are inductive and best for generating new theories, and quantitative methods are deductive and build on previously identified theories. Both are important for the scientific enterprise. As we discussed in Chapter 2, a theory is actually an explanation for a given phenomenon well supported by evidence.

Between biology and psychology, we observe the animosity between the "hard" and "soft" sciences. There is a pervasive attitude that is actually quite old (based on my research, I saw that it was first reported in the early 20th Century) that social sciences are not as rigorous as the hard sciences and are somehow "lesser." This idea stems from the difficulties associated with human subjects research. It is difficult to fully "control" for all variables that could influence a phenomenon or outcome. This harkens back to Chapter 2 and Chapter 13. In Chapter 2, we learned about how science does not follow one strict method. There are many appropriate ways to do rigorous, quality scientific research. In Chapter 13, we discussed the difficulties associated with human subjects research. In biology research, model organisms (which we discussed in Chapter 11) can be utilized to decrease variability between experiments. This doesn't work in the real world though. This is why clinical trials are so extensive (and as discussed in Chapter 13, expensive) leading up to approval by the Food and Drug Administration (FDA) and why there is still monitoring after FDA approval. You can't inbreed a population of humans to minimize variability in research. Not only is this unethical, it doesn't adequately represent the humans a treatment is intended to address. It is difficult to study humans, and huge amounts of evidence are required to make any claims.

We see these same issues in classroom research, too. Learners (and their teachers) bring with them many different attitudes, perspectives, opinions, and ways of thinking into the classroom.

Humans are variable and messy, and consequently social sciences research has to account for that. An interesting observation I've made over the last few years is that in spite of how "hard" scientists poo-poo social sciences research, the rigor of methods in some social sciences studies, particularly in developmental and cognitive psychology studies, is quite high. There are still issues with reproducibility (or the ability to repeat results) in the social sciences[38] ... but closer examination of the wider scientific literature would indicate this is a far more pervasive problem throughout all science disciplines.[39] In fact, getting to a theme throughout this book, namely the lack of certainty of science information (Chapters 2 and 11) and the variability of opinions, some scientists think that there is not actually a reproducibility crisis at all.[40] Essentially, there are *many* different ways of doing scientific research, and much of what we know is a dialogue around the existing evidence. Interestingly, practicing scientists struggle with this concept as well.

If there are these challenges to interdisciplinary education research, why bother going the more challenging route? Why bother with these educational reforms? How are standards and changes in pedagogy really going to influence science literacy and the confidence to engage in science issues long after students have left the classroom?

---

38 Open Science Collaboration. (2015). Estimating the reproducibility of psychological science. *Science, 349*(6251), aac4716. Doi: 10.1126/science.aac4716

39 Ioannidis JPA (2005) Why Most Published Research Findings Are False. PLoS Med 2(8): e124. https://doi.org/10.1371/journal.pmed.0020124

40 Fanelli, D. (2018). Opinion: Is science really facing a reproducibility crisis, and do we need it to?. *Proceedings of the National Academy of Sciences, 115*(11), 2628-2631.

Importance of Quality Biology Education: Why Does it Matter?

While writing this book, I attended a seminar by Dr. Brian Donovan from the Biological Sciences Curriculum Study (BSCS). Dr. Donovan's talk covered how standard genetics education in the United States can foster white supremacist leanings.[41] In the 1940s, genetic essentialism was common in biology textbooks. Genetic essentialism is a type of cognitive bias (we talked about the influence of bias and evidence evaluation in Chapter 11) and occurs when people place too much weight on how much genes influence phenotype. As discussed in Chapter 6, this could make some sense given outdated concepts of genetics, namely that one gene leads to one phenotype. However, this view is incompatible with modern genetics, especially given what we know about the role of environment on phenotype and how many genes can work in combination with each other and the environment to produce a specific phenotype. Said otherwise, the amount of pigment is in your skin is linked to genes, but who you are and what race you are and what "race" means is a much more complex question and involves more than single genes, and more than even our basic biology, but culture as well.

Biology curriculum that focuses on Mendelian inheritance and does not get into more realistic portrayals of gene expression can unintentionally foster inappropriate ideas about race. Think about it—in your high school biology classes, did you talk about sickle cell disease in African-American populations? Or Tay-Sachs disease in Ashkenazi Jews? Focusing on teaching one gene, one phenotype

---

[41] If you'd like to read more about Dr. Donovan's work, see:
Harmon, A. (2019, December 7). Can Biology Class Reduce Racism? New York Times.

(which as mentioned in Chapter 6 is the exception, not the norm), then using examples from specific racial or cultural groups encourages genetic essentialist viewpoints. Dr. Donovan's recent work is promoting educational reform involving interventions that encourage students to critique genetics essentialist models and build better models. The data suggests that the standard genetics curriculum increases acceptance of genetics essentialist thinking among high school students, but the intervention decreases adherence of genetics essentialist thinking. In other words, the standard genetics curriculum could be fueling white supremacist viewpoints by presenting incorrect and outdated models of how human genetics works.

Remember when we talked about randomized control trials in Chapter 13 as the gold standard in human-subjects research? Dr. Donovan also used this method in his studies of genetics education. I picked this example of biology education research to showcase here, because it draws on several different ideas covered in this book, while also highlighting why paying attention to the biology in our daily lives and how we talk about it is very important. His research is clearly interdisciplinary drawing from a deep understanding of biology content (the modern understanding of how human genetics works) and what we know about human bias (which we touched on in Chapter 11), plus using rigorous methods to get there. Perhaps most importantly, it illustrates why we need quality biology education research not only so students are learning content that is in line with the current understanding of biology, but because standard ways of teaching biology can actually be engendering bigger problems in society, in the case of this example fueling (even if subconsciously) white supremacist agendas.

Promoting quality biology education and an understanding of biological principles is important not only for fostering a scientifically literate society, but as an avenue for addressing racial disparities in our country. Standard approaches to teaching genetics can send the one gene, one phenotype message that some races or ethnic groups are "lesser." Almost every biology textbook and classroom will use Sickle-Cell anemia among African-Americans example when explaining genetics. Along with bias toward in- and out-groups (which we discussed in Chapter 11), it's easy for someone to look at these examples that connect a specific disease to a group "different" from them and begin to make subconscious associations between different groups of humans ... again, I can't emphasize enough that this can often occur at a subconscious level. That is part of why it is important to address this. It starts with an outdated and incorrect idea about biology that may then turn into an internalized message (because I'm not ... [insert racial or ethnic group here] ... I'm not good at [stereotypically expected gift] or I have [a vice associated with said racial or ethnic group]), and that message is then either consciously or subconsciously acted on. This phenomenon is well studied in educational research and is called stereotype threat. This is also why when conducting educational research, it is best to collect demographic information at the end of the study. Otherwise, these stereotypes can crop up in the subject participant and influence the results.

It's Not Just the Pedagogy: Attitudes are Important, Too.

Quality biology education is not solely about what the content is and how that content is presented to students, but also about relevance and the fostering of pro-science attitudes. Nothing makes me happier as an instructor when I have an excited student come up to me and tell me about this thing they saw on Netflix that reminded them of something we talked about in class or a connection they made between their own interests and what they learned in class. At the end of the day, I care far more about these pro-biology attitudes than content knowledge. Generally speaking, we all have (fairly easy) access to the internet. If someone *really* needs to know the definition of a habitat, they can quickly look it up. But what about the *confidence* and *interest* to look those things up in the first place? The confidence will last much longer than the ability to recall a definition.

Again, this comes back full circle to my motivation for writing this book. My intent is that by facilitating your connection to the rich biological world around you, you will find that learning and engaging with biology content will feel less intimidating. If it feels less intimidating, I hope that you also will feel more confident to engage with biology and make informed decisions. I've mentioned previously in this book that my goal is not to tell you how to vote or what to do ... but instead to empower you to make informed biological decisions on what you think is best both for you and for society. As mentioned in Chapter 9, we are all interconnected, and decisions about conservation, landfill space, climate change, vaccines, and genetic technologies influence us each individually and as members of society.

Closing Thoughts.

Why biology everywhere? My philosophy (one that is evidence-based) is that once students begin to make the connection between the biology classroom and the real world, the content begins to feel less abstract and scary. My hope is that you, the reader (as well as students), make connections on your own and build confidence in your ability to understand and engage with the biology all around you. Then as you go about your life, you may stop and say, "Oh! I remember reading/learning about that and why it's important." Or maybe you think twice before buying 23andMe kits for your family as a gift because maybe it isn't a good idea to share your genetic information with the company and because understanding the connection between our genes and phenotype is much more complicated than originally thought. Or maybe you give a stronger review of the evidence you use when deciding whether or not to purchase foods with labels showing they are produced with genetic engineering. Maybe you think twice about whether the arts are really separate from the sciences and consider sending your kids to a STEAM rather than a STEM school or complain a little less about the cost of your next prescription. Maybe you vote to fund arts programs or start engaging in curbside composting in your community. Whatever it is, my hope is that the relevance of biology to your daily life is both more accessible and relatable than before you picked up this book.

How else does biology influence your life? What about the other sciences? Here is a challenge for you. Can you count how many different ways you interact with biology over the course of your day? Even the process of naturally waking up in the morning when the sun hits your face is a biological process and sets off a string of chemical reactions. Throwing off these reactions (or your circadian rhythm) is also why jet lag is so unpleasant. Looking out your window, how many different organisms do you see and are they interacting with one another? My son and I start every day looking out the window and pointing what we see. This morning we saw two squirrels chasing each other around a tree (a within species interaction). Throughout this book, I've given a wide variety of examples of where I've seen biology content leak from the classroom into my daily life. What can you relate to?

Will you approach anything differently now that you've read this book? What connections have you made while reading, and did anything surprise you? Keep the conversation going and be sure to follow me on social media for more *Biology Everywhere* content.

Index

A
Adenosine triphosphate 55-58
Adhesion 28
Adolescence 146-148
Agar Art 174-175
Alcoholic Fermentation 57-59
Apoptosis 41, 45
Attachment 142-143
Authentic Science Inquiry 162
Autotroph 50

B
Basement membrane 44
Basic research 188-189
Belmont Report 194
Bias 16-19, 152-153
Bioaccumulation 52, 138
Biodiversity 96, 107-113
Biodiversity hotspots 109-110, 113
Biological illustration 172-176
Biosphere 117
Bisphenol Alcohol 32, 35
Blastocyst 134
Blood type 70-72

BRCA 44-45, 83
Breastfeeding 18-19
Bushmeat crisis 122-123

C
Cancer 40, 42-47, 81
Carbohydrate 32-33
Carcinogen 47
Celiac Disease 80-81
Cell signaling 40-42
Cell theory 23, 39
Cell 37-48
Cellular respiration 55-57
Certainty of science knowledge 155-158
Chemical 25-29
Chemistry 3, 25-35
Childhood 143-146
Cholesterol 34
Chromosome 64
Climate change 50
Clinical trial 188, 195-199
Codominance 70
Cohesion 28
Community 117
Compost 127-130
Conception 134
Conservation 121
Conservation task 146
Control group 12
Creativity 19-21
CRISPR 87, 91

D

De-extinction 90-92
Deterministic genetics 82-83
Developmental biology 133, 173
Diabetes 53
Dihydrogen monoxide 27
DNA 45
Dihydrogen monoxide 27
DNA 62-67
DNA damage 45
Dominant genes 69
Dopamine 179
Double-blind 13, 196
Dutch Hunger Winter 75

E

Ecology 3, 117-121
Ecosystem 117
Electron transport chain 56-57
Emerging adulthood 148
Energy 51-53
Epigenetics 22, 68, 74-76, 138-139
Epistemological beliefs about science 154-155
Epistemic motive 160
Evidence 7
Evolution 3, 96-107, 113
Executive function 144
Exercise 54
Extinction 90, 106

F

False belief task 143-144

Fat 32-34
Fear conditioning 76
Fermentation 57-59
Firsthand bias 156, 166-167
Fossil 101
Fraternal twins 68

G
Galileo Galilei 21-22, 156
Gametes 67
Gastrulation 135
Gene 64-65
Gene sequencing 80-85
Genetic diversity 66-67, 112
Genetic essentialism 209
Genetic modification 85-88
Genetics 61-77
Genotype 69-72
Germ theory 8
GFP 20
Gleevac 31
Glucagon 53
Glucose 53-55
Glycolysis 55-57
Grants 190-191
Gregor Mendel 69

H
Habitat 92, 118
Hardin's Tragedy of the Commons 130-131
HeLa 83
Hox genes 136

Human subjects research 193-195
Hydrogen bond 28
Hypothesis 23

I
Identical twins 68
Implantation 134-135
In vitro fertilization 89
In-group 165-166
Incomplete dominance 70
Institutional review board 194-195
Insulin 53
Intergenerational trauma 75-76
Invasive species 118-120

J
Jean Piaget 145-146
Justification of science knowledge 155, 161-164

K
Kreb's cycle 56-57

L
Lactic acid fermentation 57, 59
Laws 23
Linalool 29, 35, 198

M
Macromolecules 32-33
Marshmallow task 144-145
Maxwell's equations 23
Meiosis 68

Mendelian inheritance 69, 208
Metastasis 42-43
Mitochondria 56, 92-93
Mitosis 40, 67
Model organism 173
Motivated reasoning 159
Music therapy 178
Mutation 65-66, 97, 113

N
Natural selection 99
Nature of science 155
Neurotransmitter 178
Next Generation Science Standards 10, 154, 181, 203
Nucleic acid 32
Nucleotide 63, 66

O
Oncogene 46
Outgroup 165
Ovulation 134

P
p53 45
pedigree 72-73
peer-review 15, 162
phenotype 69-70
photosynthesis 49
placebo 196
placebo effect 196
placenta 135
polydactyl 136-137

population 117-121
preimplantation genetic diagnosis 88-90
producer 51
protein 32-33
protooncogene 46
Punnett square 69, 72
Pyruvate 56

Q
Qualitative research 205
Quantitative research 205

R
Randomized control trial 196-198, 209
Rare disease 192
Recessive genes 69
Reflexes 140
Regeneration 148-149
Robert Hooke 37-38
Rosalind Franklin 62-63
Roundup 86-87

S
Salicylic acid 30
Scientific method 8-13
Selective breeding 85, 99
Selective serotonin reuptake inhibitor 29
Sensitive period 139
Sex linkage 73
Sexual reproduction 67-68
Sexual selection 99-100
Simple inquiry 10, 161

Social referencing 141-142
Sonic hedgehog 136-137
Source of science knowledge 155, 164-168
Speciation 90, 106-107
Species survival plan 91, 112
STEAM 183
Structure of science knowledge 155, 164-168

T
Taxenes 30
Taxonomy 108
Teratogen 137
Theory of mind 143-144
Theory 7, 8, 12
Three-parent babies 92-94
Transgenic organism 85-86
Translational research 188-189
Tumor suppressor 45

V
Venom 31
Vestigial organ 102-103

W
Water 27-28

Z
Zygote 39, 68, 134-135

ABOUT THE AUTHOR

Dr. Melanie Peffer has a BS and PhD in molecular biology from the University of Pittsburgh. Her current research efforts focus on how people learn, understand, and engage with biology content. She regularly speaks on her biology learning research at both national and international venues. She particularly enjoys talking about biology with the general public and youth and is an in-demand speaker and educator. Dr. Peffer lives in Colorado with her husband, young son, and two cats. To hear more *Biology Everywhere* content and stay up-to-date on upcoming *Biology Everywhere* events, visit www.biologyeverywhere.com.

www.ingramcontent.com/pod-product-compliance
Lightning Source LLC
Chambersburg PA
CBHW071231080526
44587CB00013BA/1565